MASTERING OHIO'S GRADE 5 SCIENCE ACHIEVEMENT TEST

MARK JARRETT

Ph.D., Stanford University

STUART ZIMMER

JAMES KILLORAN

JARRETT PUBLISHING COMPANY

EAST COAST OFFICE
P.O. Box 1460
Ronkonkoma, NY 11779
631-981-4248

SOUTHERN OFFICE
50 Nettles Boulevard
Jensen Beach, FL 34957
800-859-7679

WEST COAST OFFICE
10 Folin Lane
Lafayette, CA 94549
925-906-9742

www.jarrettpub.com
1-800-859-7679   Fax: 631-588-4722

Jarrett Publishing Company
Post Office Box 1460
Ronkonkoma, New York 11779

ISBN 1-882422-99-6
Printed in the United States of America
First Edition
10 9 8 7 6 5 4 3   10 09 08

ACKNOWLEDGMENTS

The authors would like to thank the following Ohio educators who helped review the manuscript. Their respective comments, suggestions, and recommendations have proved invaluable in preparing this book.

Susan Clay
2005–2006 President, Science
Education Council of Ohio (SECO)
Parma, Ohio

David Shellhaas
Science Curriculum Specialist
Darke County Educational Service Center
Greenville, Ohio

Jody Shepherd
Teacher, Berwick Elementary School
Columbus Public Schools
Columbus, Ohio

Missy Zender, Ph.D
Science Curriculum Specialist
Summit County Educational Service Center
Cuyahoga Falls, Ohio

Layout, graphics, and typesetting: Burmar Technical Corporation, Albertson, N.Y.

This book is dedicated…

to my wife, Gośka, and my children Alexander and Julia — *Mark Jarrett*

to my wife Joan, my children Todd and Ronald, and
my grandchildren Jared and Katie Rose — *Stuart Zimmer*

to my wife Donna, my children Christian, Carrie, and Jesse,
and my grandchildren Aiden, Christian, and Olivia — *James Killoran*

TABLE OF CONTENTS

UNIT 1

AN INTRODUCTION TO THE OHIO GRADE 5 SCIENCE ACHIEVEMENT TEST

This year you will take the **Grade 5 Science Achievement Test**. Everyone wants to get a high score on this test. Unfortunately, just wanting a high score is not enough. You will really have to work at it! With this book as your guide, you should be better prepared for the test — and even enjoy studying for it.

WHAT WILL BE ON THE TEST?

Let's start by learning about the test itself. The **Grade 5 Science Achievement Test** has a total of **48 points**. There are three different types of questions:

| Multiple-Choice Questions | Short-Answer Questions | Extended-Response Questions |

MULTIPLE-CHOICE QUESTIONS
The test has 32 multiple-choice questions that count towards your score. Each multiple-choice question will have four choices and be worth **1 point**. There will be four additional practice questions that **do not** count towards your score.

SHORT-ANSWER QUESTIONS
Short-answer questions require a brief response in which you write a few sentences to answer the question. The test will have four short-answer questions that count towards your final score. Each question will be worth **2 points**.

EXTENDED-RESPONSE QUESTIONS
An extended-response question requires a longer answer. It usually has more parts to it than a short-answer question. There will be two extended-response questions that count towards your score. Each will be worth **4 points**.

LESSON 1

OHIO'S SCIENCE STANDARDS

The **Grade 5 Science Achievement Test** covers six science standards that you have studied in the third, fourth, and fifth grades:

Each standard has several "benchmarks" you should know. The questions on the test will focus on these benchmarks. The chart below shows you how many questions will be asked about each standard on the test. It does not include the four additional practice questions that will not count towards your final score.

Standard	Multiple-Choice Questions (1 point each)	Short-Answer Questions (2 points each)	Extended-Response Questions (4 points each)	Total Points
Scientific Ways of Knowing; Scientific Inquiry; Science and Technology	4 to 6	1 or 2	1*	10 to 14
Earth and Space Sciences	8 to 10			10 to 14
Physical Sciences	8 to 10	2 or 3	1*	10 to 14
Life Sciences	8 to 10			10 to 14
Total Number of Items (38)	32	4	2	38

On any test, only 1 of these 3 standards will have an extended-response item that counts.

As you can see, the **Ohio Grade 5 Science Achievement Test** is a balanced test. It tests all three of the main fields of science equally. It also gives the same weight to the three standards about the general nature of science.

HOW THIS BOOK CAN HELP YOU

This book provides a complete "refresher" of the knowledge and skills you will need to do your best on the **Grade 5 Science Achievement Test.**

TOOLS FOR MASTERING THE TEST

In **Lesson 2** of this book, you will explore each of the four types of multiple-choice questions found on the test. **Lesson 3** will show you how to answer short-answer and extended-response questions. You will also examine sample questions, sample responses, and learn a step-by-step approach for answering these.

A REVIEW OF THE BENCHMARKS

The main part of the book consists of four units. Each of these units examines a different Ohio science standard and its benchmarks.

UNIT 2: THINKING LIKE A SCIENTIST

This unit consists of several lessons dealing with *Scientific Ways of Knowing, Scientific Inquiry* and *Science and Technology.* In this unit, you will learn how scientists think and approach problems. You will also learn about laboratory safety, measuring, and reading graphs and charts. What you learn here applies to all fields of science.

Students work together on a class experiment while employing safety practices.

UNIT 3: LIFE SCIENCES

In this unit you will learn about living things. You will learn about plants and animals and how they differ. You will also learn about plant and animal life cycles, and how different living things interact in ecosystems.

UNIT 4: PHYSICAL SCIENCES

In this unit, you will learn about matter, its different forms and physical properties, and the difference between physical and chemical changes. You will also learn about motion, force, and energy.

UNIT 5: EARTH AND SPACE SCIENCES

This unit reviews what you need to know about the ***Earth and Space Sciences***. The unit explores the processes that shape Earth's surface, Earth's resources and the weather. You will also learn about the sun, moon, solar system, and stars.

THE ORGANIZATION OF THE BOOK

Each content unit of this book has the following features:

★ Each lesson opens with a list of *Major Ideas*, highlighting the most important concepts in the lesson.

★ Each lesson is divided into smaller sections to help you understand a major science topic.

★ *Applying What You Have Learned* activities help you to recall and apply what you have just learned.

★ Each lesson includes several *Study Cards* to help you review the most important *facts*, *concepts*, and *relationships* in that lesson.

★ Each lesson finishes with a *What You Should Know* box, summarizing the major ideas found in that lesson.

★ Each lesson provides test-style practice questions at the end. Each question is identified by its Ohio Standard, Benchmark, and Grade Level Indicator.

★ Each unit includes a *Concept Map*, which visually summarizes the most important information for that standard.

★ Each unit concludes with a checklist of Ohio Science Benchmarks. These checklists are designed to make sure you have mastered each benchmark before you move on to the next unit.

FINAL PRACTICE TEST

The last part of the book consists of a *complete* practice test in science, just like the actual **Grade 5 Science Achievement Test**.

LESSON 2

HOW TO ANSWER MULTIPLE-CHOICE QUESTIONS

There will be four types of multiple-choice questions on the **Grade 5 Science Achievement Test**. These questions will test your ability to:

Recall scientific facts, concepts and relationships	Understand and analyze scientific information	Demonstrate investigative processes of science	Apply concepts and make connections with science

This lesson will help you to recognize each type of question. You will also learn a method to help you answer each type of question on the test. Let's begin by looking at the first type of question.

RECALLING INFORMATION

Many questions on the test will simply ask you to *recall* or *identify* scientific facts, concepts and relationships. For example, examine the question below:

> **1.** In some ways, Earth is like the moon. In which way is Earth different?
> A. Earth has both night and day.
> B. Earth has gravity, which attracts objects.
> C. Earth is surrounded by a blanket of air.
> D. Earth's surface is covered by vast craters.

As you can see, this question tests your ability to recall information about Earth and the moon.

UNLOCKING THE ANSWER

🔑 What do you think is the answer to Question 1? _____

🔑 Explain why you selected that answer. _____

_____

USING THE E-R-A APPROACH

Whatever type of question you are asked, we suggest you follow the same three-step approach to answer it. Think of this as the "**E-R-A**" approach:

EXAMINE the question

RECALL what you know

APPLY what you know

Let's look at each of these steps to see how they can help you to select the right answer.

STEP 1: EXAMINE THE QUESTION

Start by reading the question carefully. Make sure you understand any information the question provides. Then make sure you understand what the question is asking for.

HINT This question asks you to identify a difference between Earth and the moon — "*In which way is Earth different?*"

STEP 2: RECALL WHAT YOU KNOW

Next, you need to identify the topic of science that the question asks about. Take a moment to think of what you know about that topic. Mentally review the important concepts, facts, and relationships you can remember.

In this case, you should think about what you can recall from your study of space. The question asks how Earth differs from the moon. Sit back and think about what you recall about this topic.

★ *You may remember that Earth has water, an atmosphere, and different life forms.*

★ *In contrast, the moon has no water, atmosphere, or life. Its surface is covered by craters.*

★ *You should also recall that both Earth and the moon have gravity as well as day and night.*

STEP 3: APPLY WHAT YOU KNOW

Finally, take what you can recall about the topic and apply your knowledge to answer the question. Sometimes, it helps to try to answer the question on your own **before** you look at the four answer choices. Then, look to see if any of the answer choices is what you think the correct answer should be. Next, review all of the answer choices to make sure you have identified the best one. Eliminate any answer choices that are illogical or obviously wrong. Then select your final answer.

*Here the question asks you to identify one way Earth is different from the moon. To answer this question, you need to recall the characteristics of both Earth and the moon. Then you have to apply this information by selecting one characteristic that applies to Earth but **not** to the moon.*

*Start by eliminating any choice that says something true of **both** Earth and the moon. Also eliminate any choice that says something true of the moon but not Earth. The following choices are therefore wrong:*

★ **Choice A:** *"has night and day" is true of both Earth and the moon.*

★ **Choice B:** *"has gravity" is also true of both.*

★ **Choice C:** *"is surrounded by a blanket of air" is the only answer choice that says something about Earth that is **not** also true about the moon. Earth is covered by a blanket of air, known as the "atmosphere." You should recall that the moon has no atmosphere.*

★ **Choice D:** *"is covered by vast craters" is true of the moon but **not** Earth.*

Therefore, Choice C is the correct answer.

ANALYZING SCIENTIFIC INFORMATION

A second type of question found on the test will examine your ability to *analyze scientific information*. To *analyze* means to break something up into its various parts in order to understand it better. Very often, this kind of question will present you with scientific data or observations and ask you questions about them.

Such analysis questions may ask you to:

Organize, summarize or evaluate information

Make estimates

Choose the best model to represent the results of an experiment

Draw conclusions from information

Describe patterns or relationships in observations or data

Let's look at a sample question asking you to analyze scientific information.

2. When copper wires with plastic coating are connected to a battery and an electric bulb, as shown in Figure 1, the bulb lights up. When one of the wires is cut, as shown in Figure 2, the light goes out.

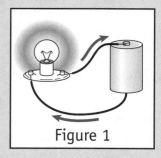

Figure 1

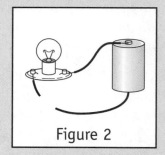

Figure 2

Why does the light go out?

A. The bulb is too overheated.
B. Electricity needs a complete circuit to flow.
C. Cutting the wire drains the battery of electricity.
D. Electrical energy always changes into light energy.

UNLOCKING THE ANSWER

🔑 How does *Figure 2* differ from *Figure 1*?_____

_____

_____

🔑 What is the answer to Question 2? _____

🔑 Explain why you selected that answer. _____

_____

As you can see, this question includes two different diagrams. These diagrams show two ways wires can be connected to a battery and a light bulb. In ***Figure 1***, the bulb is lit up. In ***Figure 2***, the light has gone out. The question then asks you why the light in ***Figure 2*** has gone out.

How should you answer a question that asks you to analyze scientific information? Again, you should try the "E-R-A" approach. Let's see how this approach could be used to answer this question.

USING THE "E-R-A" APPROACH

✦ Step 1: E̲XAMINE the Question

Read the question carefully. Be sure to examine any data or observations it may include. Usually, data will be presented as a diagram, graph, or table. Then be sure you understand what the question is asking for.

In question 2, the data is presented in two diagrams. The question asks you to explain why the light in Figure 2 has gone out. Think how the diagrams are different.

✦ Step 2: R̲ECALL What You Know

Next, take a moment to think about any important facts, concepts and relationships you can recall about the subject of the question.

*Here, the question asks about **electricity**. Think of what you can remember about this topic. You might recall that electricity can create heat and light energy as well as magnetic force. You might also remember that electricity can flow in a circuit or path. However, for electricity to flow in a circuit, the circuit must be complete.*

USING THE "E-R-A" APPROACH

✦ **Step 3: APPLY What You Know**

Apply the information that you remember to select the correct answer. You might first attempt to answer the question without looking at the choices. Then examine the answer choices and eliminate those you know are wrong. Finally, select the best answer from the remaining answer choices.

You should be able to determine that the bulb in Figure 2 will not light up because the circuit is incomplete. You should be able to quickly eliminate some of the choices:

⭐ *Choice C makes no sense. Cutting the wire does not drain the battery of electricity. If the wire were reconnected, the light would light up again.*

⭐ *Choice D correctly describes what happens when the wires are connected. As electricity flows through the thin wires in the light bulb, some of this electrical energy changes into light energy. However, this does not explain why the light goes out when the wire is cut.*

⭐ *Choices A and B remain. Since the wire is cut, the circuit is incomplete. No electricity will flow through the circuit or into the light bulb.* **Choice B** *is the correct answer.*

EXPLAINING SCIENTIFIC INVESTIGATION

Some questions on the **Grade 5 Science Achievement Test** will examine your ability to think like a scientist. Here are some of the things you may be asked to do:

| Make observations | Describe the procedures used in an experiment | Make scientific measurements | Make predictions or develop questions based on a scientific experiment |

For example, look at the following question about scientific investigation:

3. Some students want to learn about changes in the weather.

   Which instrument would help them to measure air pressure?

   A. telescope
   B. barometer
   C. thermometer
   D. graduated cylinder

UNLOCKING THE ANSWER

What do you recall about what each of these instruments measures?

_____

What do you think is the answer to Question 3?_____

Explain why you selected that answer. _____

_____

USING THE "E-R-A" APPROACH

✦ Step 1: EXAMINE the Question

First, examine any information in the question. Then determine what the question is asking for.

In question 3, you are told that some students are investigating the weather. The question then asks you to determine which instrument they can use to measure air pressure.

✦ Step 2: RECALL What You Know

Now, identify the topic of the question and recall those concepts and facts you know about it.

For this question, think about what you remember about weather. You might recall that air pressure is the force created by the weight of air pushing down. You might also remember what instrument scientists use to measure air pressure.

✦ Step 3: APPLY What You Know

Finally, apply your knowledge to the question. Eliminate any wrong answer choices. Select the best answer choice remaining.

If you recall that a barometer is used to measure air pressure, then you can answer the question easily. Even if you don't remember what a barometer is, you may still be able to select the correct answer. Start by trying to eliminate those choices that you know are wrong. You might recall that a thermometer measures temperature. A graduated cylinder measures volume, while a telescope observes distant objects. Since none of these are used to measure air pressure, the correct answer must be **Choice B.**

APPLYING SCIENTIFIC CONCEPTS TO "REAL WORLD" SITUATIONS

Some questions on the test will ask you to apply scientific concepts to "real world" situations. Questions dealing with "real world" situations usually ask you to:

| Apply your scientific knowledge to new situations | Use scientific concepts to solve problems | Determine which scientific procedures to use in an investigation |

Let's look at a sample question that asks you to make connections between science and the "real world."

4. Students often slip on the wet pavement in front of their school when it is raining. How can the school make the pavement less slippery?
   A. smooth any roughness in the pavement
   B. throw sand and gravel on the pavement
   C. wash the pavement with a strong detergent
   D. ask all of their students to wear smooth slippers

UNLOCKING THE ANSWER

🗝 What do you think is the answer to Question 4? _____

🗝 Explain why you selected that answer. _____

_____

_____

_____

To answer questions asking you to apply scientific concepts to solve a problem, you should again try using the "**E-R-A**" approach:

USING THE "E-R-A" APPROACH

✦ **Step 1: EXAMINE the Question**

Examine the question carefully. Try to be sure that you understand the "real world" situation that is being presented in the question.

In this question, students are slipping on the wet pavement in front of school. What can the school do to stop this?

✦ **Step 2: RECALL What You Know**

This kind of question asks you to apply your scientific knowledge to "real world" situations. You need to identify which scientific facts, concepts, and relationships best apply to the situation.

To answer this question, you need to recall what you know about the topic of force and motion. What force causes the students to slip? What force could be used to slow down or stop this unwanted motion? You may recall that friction acts to decrease motion. Friction increases if the objects rubbing against each other are rough rather than smooth.

✦ **Step 3: APPLY What You Know**

Now use what you know to answer the question. First, think how you might answer the question without looking at the answer choices. Then study the choices carefully, eliminating those that are obviously wrong. Finally, select the best answer.

In this example:

★ *Choice A would make the pavement smoother, reducing friction and making the pavement more slippery.*

★ *Choice C, cleaning the pavement, might also make the pavement more slippery.*

★ *Choice D, wearing smooth slippers would reduce friction and make the pavement even more slippery.*

★ *Choice B is the best answer. By throwing sand and gravel on the wet pavement, the school would make its surface rougher. It would become harder to slip on this surface because of the increased friction.*

Now use the **"E-R-A"** approach to answer a second question on applying scientific concepts to "real-world" situations.

5. People living in the southeastern United States sometimes face hurricanes. What step might help scientists to predict hurricanes?
   A. take measurements of the ocean's depth
   B. study the mountains along the ocean floor
   C. study the soils of the southeastern United States
   D. observe and monitor the ocean's surface temperatures

USING THE "E-R-A" APPROACH

◆ **Step 1: EXAMINE the Question**

What does the question ask you to do? _____

_____

◆ **Step 2: RECALL What You Know**

What can you recall about this topic? _____

_____

_____

_____

_____

◆ **Step 3: APPLY What You Know**

Using what you know, how does that help you to arrive at the answer? _____

_____

_____

_____

_____

In this lesson, you learned how to answer different types of multiple-choice questions. In the next lesson, you will learn how to answer short-answer and extended-response questions.

LESSON 3

HOW TO ANSWER SHORT AND EXTENDED-RESPONSE QUESTIONS

Some questions on the **Grade 5 Science Achievement Test** will ask you to write the answer in your own words.

SHORT-ANSWER QUESTIONS

There will be four short-answer questions on the test. A **short-answer question** requires you to give two facts or pieces of information. Each question is worth two points. These short-answer questions will test each of the following standards:

Two short-answer questions will test two of the following:

Earth and Space Sciences	Life Sciences	Physical Sciences

Two other short-answer questions will test two of the following:

Scientific Inquiry	Scientific Ways of Knowing	Science and Technology

EXTENDED-RESPONSE QUESTIONS

There will also be two extended-response questions on the test. An **extended-response question** will require you to give four facts or pieces of information. Each of these questions will be worth four points.

One extended-response question will test one of the following:

Earth and Space Sciences	Life Sciences	Physical Sciences

A second extended-response question will test one of the following:

Scientific Inquiry	Scientific Ways of Knowing	Science and Technology

RESPONDING TO A SHORT-ANSWER OR EXTENDED-RESPONSE QUESTION

There are many ways to approach short-answer and extended-response questions. One of the best ways is to use three main steps:

Analyze and Plan → Write Your Answer → Review and Revise Your Answer

STEP 1: ANALYZE AND PLAN

To answer either a short-answer or extended-response question, first look carefully at the directions of the question. The exact instructions for what you are supposed to do will usually be found in the "**action words**" of the question:

SOME OF THE MOST COMMON "ACTION WORDS"

Compare	To identify similarities and differences between two or more things—how they are alike and how they are not alike.
Describe	To give the characteristics of something, or tell how something changes over time.
Explain	To *explain how* something happened, *why it happened* or to *explain its effects*: • To *explain how*, tell the way in which it took place. • To *explain why*, give the reasons why it happened. • To *explain effects*, identify and describe each effect.
Identify	To name something or to tell what it is.
Predict	To try to tell what will happen in the future.
Support	To give facts or examples to back up a conclusion or point of view.
Draw	To create a diagram, map, or illustration of something.

After you have studied the "action words" of the question, you should next identify **all** of the parts of the question. Don't just rush into answering the question. Take a few moments to plan your answer. You can do this by simply jotting down a few notes you think might be helpful.

USING AN ANSWER BOX

It often helps to plan your answer with an **answer box**. This box divides up the different parts of the question. You can fill in the answer box with your ideas or simply check off each part of the box as you answer it. The answer box serves as a checklist, making sure that you answer each part of the question. Even if you decide not to write out the answer box, you should still complete this process in your head:

★ **Short-answer questions** will have **two** parts. For example, a short-answer question might ask you to *identify* *two characteristics of Earth*. Here is what an answer box might look like for this question:

First Characteristic of Earth	Your Response
Second Characteristic of Earth	Your Response

★ **Extended-response questions** will usually have **four** parts. For example, an extended-response question may ask the following:

> Many different processes help shape Earth's surface.
>
> In your **Answer Document**, identify two processes that help shape Earth's surface. For each process you identify, describe one type of evidence scientists use to investigate this process. (4 points).

Here is what an answer box might look like for this question:

FIRST PROCESS		SECOND PROCESS	
Identify	Your Response	*Identify*	Your Response
Describe Evidence	Your Response	*Describe Evidence*	Your Response

Look at the following short-answer question. In the space below, create an answer box that could be used to answer the question. You do not have to fill in the boxes you create for your response. You will learn more about this topic in **Lesson 10**.

Animals and plants often bring about changes to their environment.

In your **Answer Document**, identify two examples of changes that living things have brought to their environment. (2 points)

STEP 2: WRITE YOUR ANSWER

The next step in responding to a short-answer or extended-response question is to *write* your answer.

★ You can use the notes you created in your answer box to write your answer.

★ It may help to "echo" or restate the question to introduce your answer. To echo the question, repeat it in the form of a positive statement. For example, suppose you were responding to the short-answer question on the top of this page. If you were to echo this question, your answer might begin:

Living things bring many changes to their environment.

★ Finally, turn each point in your notes or answer box into one or more complete sentences. Check off sections of your answer box each time you complete that part of your answer.

STEP 3: REVIEW AND REVISE YOUR ANSWER

The first person to read your answer should be **YOU** — *not* the person scoring it. Once you have finished writing, read over your answer *before* you hand it in. Make sure you have provided all of the information required by the question. As you review what you have written, ask yourself three important questions:

★ Did I follow **all of the directions** in the question?

★ Did I complete **all of the parts** of the question?

★ Did I **provide enough details, examples,** and **reasons** to support my answer?

HOW YOUR ANSWER WILL BE SCORED

To see how your answer will be scored, let's look at a model question based on Benchmark B of Earth and Space Sciences:

Many different processes help shape Earth's surface.

In your **Answer Document**, identify two processes that help shape Earth's surface. For each process you identify, describe one type of evidence scientists study to investigate that process. (4 points).

Read each of the following answers carefully. Then give each one a score of **0, 1, 2, 3,** or **4** — with **"4"** as the best score.

RESPONSE A:

Many different processes help to shape the Earth's surface. One of these processes is the folding of Earth's surface. This folding helps to build mountains. Scientists often study the rocks found in mountains to find evidence of this folding of Earth's surface. A second process affecting Earth's surface is erosion.

Your score: [　　　] Explain why you gave that score. _____

_____

_____

_____

RESPONSE B:

> Two processes that shape Earth's surface are volcanoes and erosion. Volcanic eruptions help shape Earth's surface. Volcanoes throw lava on Earth's surface. Volcanoes can even create new islands in the ocean. Scientists investigate volcanoes by studying rocks and exploring active volcanoes. Erosion also shapes Earth's surface. Wind and water can break down rock and carry sand and soil away. Scientists investigate erosion by looking at places where erosion has happened.

Your score: [   ] Explain why you gave that score. _____

_____

_____

_____

RESPONSE C:

> Freezing water shapes Earth's surface by breaking down rocks. Rivers shape Earth's surface by carrying away soil and depositing it in a new place.

Your score: [   ] Explain why you gave that score. _____

_____

_____

_____

RESPONSE D:

> Many processes shape Earth's surface. The gravity from the moon sometimes causes Earth's surface to swell. This can build mountains. Scientists study the position of the moon to see its role in making mountains. Plant growth reshapes land surfaces by breaking up rocks.

Your score: [   ] Explain why you gave that score. _____

_____

_____

_____

RESPONSE E:

> Earthquakes shape Earth's surface.

Your score: [　　　　] Explain why you gave that score. _____

_____

_____

_____

   Test scorers will use a **rubric**, or scoring guide, to score student responses on the **Grade 5 Science Achievement Test**. The rubric tells a scorer what information an extended-response should include to receive a score of **0, 1, 2, 3,** or **4 points**. Look at the rubric for the question you just scored:

EXTENDED-RESPONSE SCORING RUBRIC

Standard and Benchmark Assessed
Standard: *Earth and Space Sciences*

Benchmark: Summarize the processes that shape Earth's surface and describe evidence of those processes.

Rationale: This question asks students to identify two processes that shape Earth's surface and describe the evidence scientists study to investigate these processes.

The response will receive full credit if the response identifies two processes and describes at least one type of evidence scientists use to investigate each process.

Processes may include:

★ Erosion	★ Landslides	★ Weathering
★ Earthquakes	★ Mountain building	★ Deposition
★ Volcanic eruptions		

Scientific evidence may include:

★ **Erosion:** Scientists observe areas where erosion is taking place; they conduct experiments on the ability of wind, water, and other materials to break down and carry off rock, sand, and soil.

★ **Earthquakes:** Scientists observe earthquakes and monitor seismic waves (waves created by earthquakes).

★ **Volcanic eruptions:** Scientists observe active volcanoes and study volcanic rock.

↳ Scoring Guidelines

Points	Student Response
4	The response identifies two processes that shape Earth's surface and describes evidence used by scientists to investigate each process.
3	The response identifies two processes that shape Earth's surface and describes evidence used by scientists to investigate one of these processes.
2	The response identifies two processes that shape Earth's surface without any examples of evidence used to investigate either process OR the response provides one process and describes evidence that scientists use to study that process.
1	The response identifies one process but fails either to describe any evidence that scientists use to investigate that process or to identify a second process.
0	The response fails to demonstrate any understanding of the processes that shape Earth's surface. The response does not meet the criteria required to earn 1 point. The response indicates inadequate or no understanding of the task and / or the idea or concept needed to answer the item.

Now that you have seen the rubric for scoring responses to this question, would you change any of the scores you gave before? _____ Why or why not? _____

_____

_____

Based on the rubric above, the responses would most likely be scored as follows:

★ **Response A.** The student correctly identifies two processes that help shape Earth's surface — the folding of the Earth's crust to build mountains and erosion. However, the student only describes evidence used to investigate mountain building. The student does not describe evidence that scientists use to investigate erosion. Since only three of the four required points are found in this response, it should receive **a score of 3.**

★ **Response B.** The student correctly identifies two processes that shape Earth's surface — volcanic eruptions and erosion. The student also describes evidence scientists use to investigate each process. Since all four of the required points are found in this response, it should receive **a score of 4.**

★ **Response C.** The student actually identifies three processes that help shape Earth's surface — weathering (freezing), erosion, and deposition. However, the student cannot receive more than two points for this part of the question. The student fails to describe any type of evidence used by scientists to investigate any of these processes. Since only two of the required points are found in this response, it should receive **a score of 2**.

★ **Response D.** The student correctly identifies one process that shapes Earth's surface — plant growth, which helps to break down rocks. However, the second process that the student identifies is incorrect. The moon's gravity causes the tides, but it does not cause Earth's land forms to swell. This is not a cause of mountain building. The student also fails to describe any evidence scientists use to investigate how plant growth shapes Earth's surface. Since only one of the required points is found in this response, it should receive **a score of 1**.

★ **Response E.** The student correctly identifies one process that shapes Earth's surface — earthquakes. Although the student has written only four words, the response still manages to provide one of the required points. The response should therefore receive **a score of 1**.

As you can see from this scoring exercise, the most important part of answering any short-answer or extended-response question is:

✳ reading the question carefully
   and
✳ answering **all the parts** of the question with the **correct information**.

The length of your answer will not determine the score you receive. The number of lines you write is less important than providing correct information that fully answers the question.

WHAT YOU SHOULD KNOW

Place a check next to those boxes that you recall and understand:

☐ The **Grade 5 Science Achievement Test** will have four (4) short-answer questions and two (2) extended-response questions.

☐ Be sure to read each question carefully. Plan your answer. It is sometimes helpful to use an answer box. Next, write your answer. Finally, review and revise what you have written.

☐ Be sure to answer all parts of the question with the correct information.

UNIT 2

THINKING LIKE A SCIENTIST

The **Grade 5 Science Achievement Test** will examine your understanding of several fields of science. In this unit, you will look at some basic principles common to all these fields. To try to understand the natural world, scientists make observations, ask questions, develop theories, form and test hypotheses, and share ideas. In the next four lessons, you will learn how scientific

knowledge changes, how scientists conduct investigations, how investigation results are analyzed, and how scientific knowledge is linked to technology.

LESSON 4: SCIENTIFIC WAYS OF KNOWING

In this lesson you will learn what science is. You will also look at the methods scientists use to increase their understanding of the natural world.

LESSON 5: SCIENTIFIC INQUIRY: DESIGNING AN EXPERIMENT

This lesson looks more closely at the process of scientific investigation. You will learn how scientists form hypotheses and design experiments to test their ideas. You will also learn about the need for laboratory safety.

LESSON 6: SCIENTIFIC INQUIRY: CONDUCTING AN INVESTIGATION

In this lesson, you will learn how scientists measure length, mass, volume, and temperature. You will also learn how they organize and analyze their data. Finally, you will learn how scientists communicate their results to others.

LESSON 7: SCIENCE AND TECHNOLOGY

In this lesson, you will learn how science and technology are related.

SCIENTIFIC WAYS OF KNOWING

In this lesson, you will learn what science is. You will also learn how science increases our understanding of the natural world.

— MAJOR IDEAS —

A. **Science** is a special way of investigating and explaining the natural world. Scientific knowledge is logical. It is based on evidence and can often be used to predict future events.

B. A **fact** is a statement that can be checked to see if it is true. An **opinion** is a statement of personal belief.

C. Scientists observe nature and conduct **controlled experiments** to produce scientific evidence. They make **observations**, collect **data**, and **record** their observations and measurements.

D. Scientists **repeat** their experiments or have others repeat them. Experiments are repeated to see if they continue to produce the same results.

E. Scientists use **models** and **theories** to explain the evidence they find.

F. Men and women from all countries and backgrounds have contributed to the growth of scientific knowledge.

WHAT IS SCIENCE?

What is science? **Science** is a special way of investigating and explaining what happens in the world. Different types of scientists study every kind of natural event.

★ Some scientists study why it rains.
★ Other scientists study why children resemble their parents in different ways.
★ Still other scientists study why Earth's surface changes.
★ Scientists study the stars, the oceans, the atmosphere, and living things.

What do scientists have in common? What all these different types of scientists have in common is the way they approach their subject. They all use similar **methods of investigation** to discover useful facts. Then they all rely on these facts to develop **logical explanations** of what happens in the natural world. **Scientific explanations** are logical and are based on factual evidence.

FACT AND OPINION

Scientific knowledge is based on observable facts. Do you know what a fact is? Can you tell the difference between a fact and an opinion?

FACTS

A **fact** is a statement that can be checked to see if it is correct or true. "*Humans first landed on the moon on July 20, 1969.*" This is a statement of fact. Many people witnessed this event on television at the time.

Buzz Aldrin steps out on to the moon's surface, July 20, 1969.

Precise measurements often provide facts that scientists can use. For example, a scientist may state that at the start of an experiment, water in a beaker was 30°C. This is a fact. The scientist used a thermometer to measure the temperature of the water. Other scientists could also have measured the temperature of the water to check that it was 30°C. A scientist can repeat the same experiment, starting with a beaker of water at precisely 30°C.

OPINIONS

An **opinion** is a statement of personal feeling or belief. Words such as *think*, *feel*, and *believe* often indicate that a statement is an opinion. For example, this statement is an opinion: "*This water feels too hot.*" Different people have different views about what is hot. No one knows exactly what "hot" means. There is simply no way to check this. This type of opinion just tells us an individual's personal feeling.

Writers sometimes make statements that look like facts even though they are actually opinions. Statements with phrases like "the nicest," "the best," or "the most interesting" express opinions, not facts. For example: "Our soap cleans the best." This looks like a statement of fact, but it is really just an opinion.

There is a second kind of opinion. People give opinions about things they are currently unsure about. For example, "It might rain tonight," is an opinion. This second type of opinion can be supported by facts. Later events or discoveries may prove or disprove this kind of opinion.

APPLYING WHAT YOU HAVE LEARNED

Below are five statements from a report by a scientist. Check the boxes below to show which statements are factual and which are expressions of opinion.

	Fact	Opinion
1. Earth has one natural satellite, the moon.	☐	☐
2. Earth rotates around its axis every 24 hours.	☐	☐
3. The full moon is quite beautiful.	☐	☐
4. There is no wind on the moon.	☐	☐
5. Biology is the most interesting field of science.	☐	☐

Write a factual statement of your own: _____

_____

Write an opinion statement of your own: _____

_____

Scientific explanations are actually a special blend of opinion and fact. They give a scientist's views on the causes of a natural event. Unlike some forms of opinion, scientific explanations must always be based on facts. They are supported by observations and data that can be checked and confirmed by others.

TYPES OF SCIENTIFIC INVESTIGATIONS

There are many ways that scientists investigate the natural world.

OBSERVATION

One way scientists investigate nature is by just **observing** it. For example, a scientist may look at the different types of clouds in the sky and record their shapes. After several days, the scientist notices that when certain clouds appear, it often begins to rain. A scientist may also observe nature by using special tools. For example, with a **microscope** a scientist can observe very tiny objects, like mineral crystals or living cells.

DATA COLLECTION

Sometimes scientists just observe and describe what they see, hear, smell, or touch. At other times, they use special tools to **measure** what they observe. For example, they may use a **meter stick** to measure the length of an object. The precise measurements that scientists collect and record are often referred to as **data**.

FIELD INVESTIGATIONS VS. CONTROLLED EXPERIMENTS

A scientist may go out into the natural world to collect data. This process is known as a **field investigation**. For example, a scientist may go to a pond to observe how tadpoles grow into frogs. At other times, a scientist may design special tests that are conducted in a closed laboratory. For example, a scientist may combine different chemicals, or observe the effects of light on the growth of plants. This kind of investigation is known as a **controlled experiment**.

RECORD-KEEPING

When scientists make observations and collect data, they **record** their results. They need to record their results to provide evidence to support their conclusions.

Recording Information. Often, scientists keep their information in a notebook, log or journal. These records must accurately state what the scientists have done and the results they received. They should be written clearly so that they can be reread and understood weeks or even months later. Other scientists also need to be able to review what has been done. These other scientists can then repeat the experiment or conduct a similar field investigation to see if they reach the same results.

APPLYING WHAT YOU HAVE LEARNED

Define the terms below. This will help you to create your own scientific glossary. Add new terms of your own in your notebook as you read this book.

GLOSSARY ON THE NATURE OF SCIENCE

Scientific Term	Definition of the Term
Fact:	
Opinion:	
Observation:	
Data Collection:	
Field Investigation:	
Controlled Experiment:	
Record-keeping:	

CHOOSING THE BEST APPROACH

The type of investigation that a scientist conducts depends on the question the scientist is trying to answer.

OBSERVATION OF NATURE

Scientists may simply go into the natural world to observe events and collect data. This approach is best for exploring what happens in nature. For example, if scientists are trying to understand how tadpoles turn into frogs, they may simply observe tadpoles in a pond.

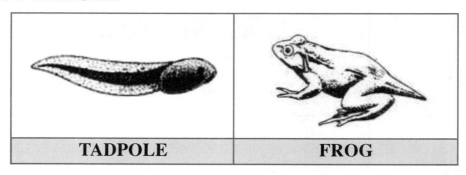

| TADPOLE | FROG |

CONTROLLED EXPERIMENT

In a **controlled experiment**, scientists actively change something to see the effect. This approach is best for seeing how two or more things are related. For example, if scientists want to see what effect a certain vitamin has on frog growth, they may collect several frogs and bring them to the laboratory. Then they will conduct a controlled experiment:

★ Some of the frogs are placed in artificial ponds *without* the vitamin.

★ Other frogs are placed in artificial ponds *with* the vitamin in the water.

The scientists will then observe what happens to both groups of frogs. The scientists will also measure the size of each frog at fixed periods, to see the effect that the vitamin is having on frog growth. You will learn more about how to conduct a controlled experiment in the next lesson.

APPLYING WHAT YOU HAVE LEARNED

Look at the list of questions on the following page. Each question is one that scientists wish to answer.

- For each question, decide whether scientists should *observe nature* or conduct a *controlled experiment*.
- Then explain why you chose that method as the best approach.

★ *How long does it take Jupiter to circle the sun?*

☐ Observation of nature   ☐ Controlled experiment

Explain your answer: _____

_____

★ *Which fertilizer best promotes the growth of bean plants?*

☐ Observation of nature   ☐ Controlled experiment

Explain your answer: _____

_____

★ *What types of rocks are found in the Appalachian Mountains?*

☐ Observation of nature   ☐ Controlled experiment

Explain your answer: _____

_____

★ What conclusions can you reach about which questions are best answered through the observation of nature (*or field investigations*) and which are best answered by a controlled experiment?

_____

_____

SCIENTIFIC EXPLANATIONS

Scientists often rely on two important tools to help them explain what happens in the natural world — models and theories.

MODELS

Scientists often use models to better understand what happens in nature. A **model** is something made or built to represent something else. It can be as simple as a diagram.

Usually a model is simpler than what it represents. For example, a model car looks like a real car, but it is much smaller. The model car may have the same shape, outside parts, and colors as the real car. Models can help scientists to explain their observations or the results of their experiments.

THE PURPOSE OF A MODEL

A good model helps scientists to see relationships and test ideas. From the model, scientists can often make predictions about what will happen. Then they can test their predictions to see if they come true. The more closely a model resembles what it represents, the better its predictions will be.

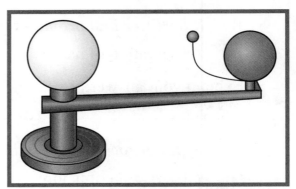

A model of the sun, moon and Earth.

The Three Gorges Dam. For example, scientists in China were about to build a very large dam on the Yangtze River to control flooding and to generate hydroelectric power. They wanted to see what effects the dam would have before it was actually built. So they built a large model of the dam. The model had the same proportions as the actual dam, but it was much smaller. The scientists used the model of the dam to conduct several experiments before actual construction began.

A model of the Three Gorges Dam, the world's largest hydropower project.

APPLYING WHAT YOU HAVE LEARNED

◆ Think of something you have learned about in science this year.

• Describe how you would make a model to show it. _____

_____

• Explain how your model makes this object or process easier to understand.

_____

_____

IMPROVING MODELS

Models are never exactly the same as the actual thing they represent. They always differ in some way — such as in size, materials, or speed of movement. Because of these differences, models can always be improved. Therefore, when you examine a model, always ask yourself:

| What is this model trying to show? | How closely does this model resemble what it represents? | How might this model be improved? |

APPLYING WHAT YOU HAVE LEARNED

✦ In a model of our solar system, a tennis ball is used to represent Earth.

- What would you use to represent the sun? _____

- What would you use to represent the planet Mercury? _____

- Where would you place these objects? _____

    _____

THEORIES

The purpose of a scientific investigation is to find out facts about the natural world. Scientists use these facts to develop **explanations** about how and why things happen in nature. If the explanation is accurate, scientists can use it to predict what will happen when the same conditions are repeated.

SCIENTIFIC THEORIES

A "big idea" in science is called a theory. A **theory** attempts to explain how and why a large number of things happen. To come up with a theory, scientists look at large amounts of data that have been collected from experiments or from observing nature. Then they try to think of a logical way of explaining all the data. For example, the ancient Greeks looked at the sky and saw that the moon, stars, and planets move. They came up with the theory that all of these objects revolved around Earth.

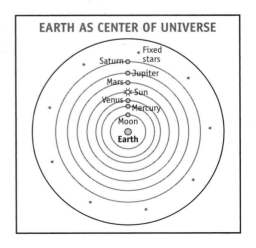

EARTH AS CENTER OF UNIVERSE

However, there were some observations the ancient Greeks could not explain. They could not explain why the planets sometimes seem to move backwards.

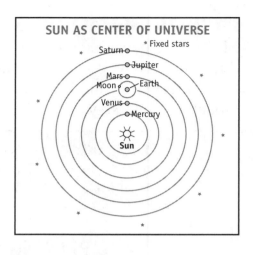

In the 1500s, Nicolaus Copernicus came up with a better theory to explain all the data. He decided that Earth moves around the sun. This new theory better explained all of the existing data.

Once scientists develop a theory, they observe nature, conduct field investigations, and design experiments to test it. If the results of repeated investigations and experiments support the theory, then the theory provides a good scientific explanation. If there are results that the theory cannot explain, then the theory has to be changed or rejected.

WHY SCIENTIFIC KNOWLEDGE CHANGES

To decide how good a scientific explanation is, scientists look at the data from their observations and experiments. No claims or conclusions can be accepted by scientists unless they are backed by data that can be confirmed. For this reason, our scientific knowledge is constantly changing and improving.

As we collect more observations and data, our ideas and conclusions change. For example, we have seen that people once believed that Earth was the center of the universe. Copernicus theorized that the sun, not Earth, was the center of the universe. With more powerful telescopes, we now realize that there are countless galaxies in the universe. The sun may be the center of our solar system, but the solar system itself is a mere speck in a much larger universe.

Copernicus

The more that scientists learn, the better they are able to improve their explanations of what we see in the world. The more that scientists learn, the better we are able to understand how things in nature really work.

APPLYING WHAT YOU HAVE LEARNED

✦ What is a "theory"? _____

_____

✦ How is a "theory" different from a "model"? _____

_____

✦ How do scientists test their theories? _____

_____

✦ Why is scientific knowledge constantly changing and improving? _____

_____

HOW PEOPLE FROM ALL CULTURES AND COUNTRIES CONTRIBUTE TO SCIENCE

Sir Isaac Newton, one of the world's greatest scientists, once said that he could see farther than others only because he stood "on the shoulders of giants." Newton did not mean that he actually stood on people's shoulders. What he meant was that science is a common human effort. Scientists depend greatly on the work of earlier scientists. What scientists know today is the result of thousands of years of human effort.

DIVERSE PEOPLE CONTRIBUTE TO SCIENCE

Men and women from every country and background have contributed to the development of science. Here are just a few examples:

★ **Aristotle** (384 B.C.–322 B.C.). In ancient times, this Greek brought together and increased the scientific knowledge of the world. He established the ideal of careful and detailed observation in science. For example, Aristotle classified all known animals and arranged them in groups.

★ **Galileo Galilei** (1564–1642). This Italian studied moving objects and was the first to use a telescope to study the planets. His investigations laid the foundations for modern experimental science.

★ **Sir Isaac Newton** (1642–1727). This Englishman discovered the laws of motion and gravity, showing that nature followed logical patterns.

★ **Louis Pasteur** (1822–1895). This French scientist's work with bacteria revealed how microscopic organisms (*germs*) cause disease. His work led to the development of pasteurization and the sterilization of medical instruments. He solved the mystery of rabies, and contributed to developing new vaccines.

Louis Pasteur

★ **Marie Curie** (1867–1934). Born in Poland, she became famous for her work on radioactivity. She discovered two radioactive elements: polonium and radium. She is the only woman to win a Nobel Prize in two different fields of science.

★ **Albert Einstein** (1879–1955). This German-Jewish scientist showed that matter could be changed to energy and that time is relative. His work laid the foundation for the development of the atomic bomb and nuclear energy.

★ **Albert Michelson** (1891–1943). Born in Poland of Jewish parents, Michelson became the first U.S. Nobel Prize winner in Physics for experiments on light. He was a professor of physics at Case School in Cleveland.

★ **Dr. Roy Punkett** (1910–1994). Educated at Ohio State University, Plunkett discovered Teflon — a coating that is extremely heat-resistant and stick-resistant.

★ **Chien-Shiung Wu** (1912–1997). Born in China, this American physicist worked on the project to develop the atomic bomb. She made important advances in the study of radiation. She is sometimes referred to as the "Madame Curie of China."

★ **Rosalind Franklin** (1920–1958). This Jewish Englishwoman was both a chemist and a biologist. She was the first to take x-ray photographs of DNA — the molecules that form the basis of all life.

★ **Stephen Hawking** (1942–present). This British physicist suffers from a disease that confines him to a wheelchair. He can only speak with the aid of a computer voice synthesizer. Despite this, he has made major contributions to our understanding of time and space.

★ **Shirley Jackson** (1948–present). This African-American physicist helped research how silicon and similar elements could conduct electricity. She served as head of the U.S. Nuclear Regulatory Commission.

Shirley Jackson

This short list of famous scientists shows how men and women from many different cultures and backgrounds have made important contributions to science.

APPLYING WHAT YOU HAVE LEARNED

People from many cultures and backgrounds have contributed to science. Select two scientists having different backgrounds. One can be from the previous page: the other should be from your own research on the Internet or in an encyclopedia. For both, identify who they are and something about their background. Describe how each made a contribution to the field of science.

Scientist	Background	Contribution to Science

CAREERS IN SCIENCE

Today, the growth of scientific knowledge continues to rely on the contributions of people from all backgrounds. There are many different careers open to those interested in science. These careers include:

★ **Medical Doctor.** A doctor uses scientific methods to help people maintain their health and to cure those who are ill.

★ **Meteorologist.** A meteorologist studies and predicts the weather.

★ **Marine Biologist.** A marine biologist studies how some plants and animals survive and interact in the oceans and in other bodies of water.

★ **Chemist.** A chemist studies different chemicals and how they interact. A **biochemist** studies the chemistry of life. Biochemists work to find new drugs to help cure various diseases.

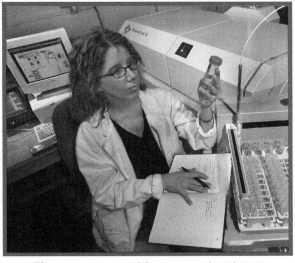

There are many exciting careers in science.

★ **Physicist.** A physicist studies motion and energy. Physicists are found working in universities, for the government, or for private companies.

★ **Science Teacher.** Many people who love science become teachers. They help their students learn how science increases our understanding of the natural world and improves our lives.

There are many other interesting careers in science to choose from, such as *microbiologist*, *astronomer*, *geneticist*, or *geologist*. Although these careers are quite different, one thing people in them all have in common is that they find science exciting. There is almost no limit to where your interest in science can take you.

APPLYING WHAT YOU HAVE LEARNED

People who enjoy science have a wide choice of careers. Identify *two* additional careers open to those interested in science. Describe what each one does.

Career	Description of that Career in Science

WHAT YOU SHOULD KNOW

☐ You should know that **science** is a special way of investigating and explaining the natural world. Science is logical, is based on evidence, and can often be used to predict future events.

☐ You should know that a **fact** is a statement that can be checked to see if it is true. An **opinion** is a personal belief that cannot be checked by others.

☐ You should know that scientists observe nature and conduct controlled experiments to produce scientific evidence. They make observations about what they see and they collect data.

☐ You should know that scientists repeat their experiments or have others repeat them to see if they continue to produce the same results.

☐ You should know that scientists use models and theories to explain the evidence they find.

☐ You should know that a diverse group of men and women have made important contributions to the growth of scientific knowledge.

☐ You should know that there is a wide range of careers in science.

CHAPTER STUDY CARDS

You will find *Study Cards* at the end of each lesson. They highlight the essential information in that lesson. You can copy or photocopy these cards. You are encouraged to make your own cards to reinforce the most important concepts and facts.

Scientific Ways of Knowing

★ **Science.** A special way of investigating and explaining what happens in nature.

★ **Fact.** A statement that can be checked to see if it is correct or true.

★ **Opinion.** An expression of personal feeling or belief that cannot be proven.

★ **Theory.** Explanation of a large amount of data, which can be tested and revised.

★ **Model.** A replica scientists use to better understand what happens in nature.

★ **Scientific Knowledge.** Our understanding improves as new data and observations are collected, and as new theories are developed.

Scientific Investigation

★ **Observation.** Scientists look at the natural world. They use tools, such as a microscope, to help them observe the natural world

★ **Data Collection.** Scientist use special tools to measure what they observe. They collect these measurements as **data**.

★ **Field Investigation.** When a scientist goes into the natural world to collect data.

★ **Controlled Experiment.** When a scientist conducts special tests in a laboratory.

★ **Record-Keeping.** Scientists write down their observations in a log or notebook, so they can be reviewed and repeated.

CHECKING YOUR UNDERSTANDING

★ The ancient Greeks believed Earth was the center of the universe and the planets moved around Earth. They could not explain why some planets appeared to move backwards.

★ To explain these movements, Copernicus later argued that the planets move around the sun, not Earth.

1. What conclusion can be drawn from these statements?

A. Scientific knowledge has not improved since ancient times.
B. Experiments are more important to science than observation.
C. Things that cannot be explained can lead to scientific advances.
D. The work of earlier scientists seldom influences later scientists.

Benchmark

SW: A
G5.1

Grade-level Indicator

This question examines your understanding of the growth of scientific knowledge. **Choices A, B,** and **D** all make statements that are incorrect. Our scientific knowledge has greatly improved since ancient times. Both observation and experiments are important, and scientists are often influenced by those who went before them. You have learned that scientists attempt to improve their explanations in order to explain things that at first are hard to understand. Copernicus developed a theory that explained why planets appeared to move backwards. Thus, the correct answer is **Choice C.**

Now try answering some additional questions on your own.

2. Why do scientists keep accurate records of their work?

    A. to make science more interesting
    B. so they can correctly keep track of expenses
    C. to meet requirements of the U.S. government
    D. so other scientists can repeat their experiments

    SW: C
    G4.4

3. A scientist reports a new discovery based on the results of an experiment. If the report is accurate, what should other scientists be able to do?

    A. design a better experiment
    B. use the discovery to make new medicines
    C. repeat the experiment to get the same results
    D. conduct a different experiment to get the same results

    SW: B
    G5.3

4. How does a model differ from a theory?

    A. A model can be checked, while a theory cannot be disproved.
    B. A model often looks at nature, while a theory generally does not.
    C. A model is based on observation, while a theory is based on data.
    D. A model is made to resemble something, while a theory tries to explain the data or observations.

    SW: B
    G5.2

5. A student did an experiment to find out if the temperature affects the rate of growth of a particular type of plant. What should the student do to accurately communicate the results of the experiment?

    A. Write a story about a favorite plant.
    B. Make a list of materials used in the experiment.
    C. Draw a picture of the equipment used in the experiment.
    D. Display all of the recorded measurements in a table or graph.

    SI: B
    G3.5

6. Which of the following statements expresses an opinion?

   A. Marie Curie discovered radium.
   B. This book is 30 centimeters long.
   C. Our universe may not last forever.
   D. The experiment ended at 12:35 p.m.

   SW: A
   G4.1

Use the following information to answer question 7.

Jupiter is the largest planet in the solar system. Saturn is the second largest planet. Some scientists believe that when these planets are viewed through a telescope, what we see is mainly their atmospheres — the mixture of gases that surrounds them.

7. Which would provide the best evidence in support of this theory?

   A. Jupiter and Saturn have a large number of moons.
   B. Jupiter and Saturn are farther from the sun than Earth
   C. Jupiter and Saturn are the two largest planets in the solar system.
   D. Unmanned satellites sent to these planets found very large atmospheres.

   SW: B
   G5.2

Use the following information to answer question 8.

A teacher takes a class on a field trip to a nearby pond. To show his class how the human ear hears sounds, the teacher drops a stone into the pond. The stone hitting the water creates small waves in the pond. When the waves hit a leaf floating at the edge of the pond, the leaf begins to move back and forth.

8. In this model, what does the leaf represent?

   A. a sound wave
   B. compressed air
   C. the human brain
   D. an ear drum

   ◆ **Examine the Question**
   ◆ **Recall What You Know**
   ◆ **Apply What You Know**

   SW: B
   G5.2

9. Scientific knowledge is based on the development of theories that try to explain the natural world.

   In your **Answer Document***, give one example of a theory.

   Then, show how that theory was later changed based on experiments or observations. (2 points)

   SW: B
   G5.2

*For your Answer Document, write your answers on a separate sheet of paper or in your notebook.

LESSON 5

SCIENTIFIC INQUIRY: DESIGNING AN EXPERIMENT

In this lesson, you will learn more about how scientific investigations are conducted. You will learn how scientists ask scientific questions, and develop and test hypotheses. In **Lesson 6**, you will learn how scientists conduct experiments and draw conclusions.

— MAJOR IDEAS —

A. **Scientific inquiry** is the process by which scientists ask questions and investigate the natural world.

B. The following steps are often used by scientists to conduct an investigation:

★ The scientist observes the world and **asks a question**.
★ The scientist develops a **hypothesis** to answer the question.
★ The scientist designs an experiment to **test** the hypothesis.
★ The scientist uses special **equipment** to carry out the experiment.
★ After an experiment, the scientist **interprets its results**.
★ The scientist **communicates** these results and conclusions to others.

C. Scientists take **safety** into account in all field and laboratory investigations.

STEPS OF A SCIENTIFIC INVESTIGATION

Scientific inquiry usually begins with observations of the natural world. Scientists then ask open-ended questions about what they observe. For example, a scientist might watch children fly paper airplanes. The scientist might then ask:

How can a paper airplane be made to fly a longer distance?

Scientists then try to answer their questions by building models, making observations and conducting experiments. Let's look at the steps that scientists often follow to design and conduct an experiment.

42

ASK A WELL-DEFINED QUESTION

A scientist begins by observing the world. Often what a scientist sees raises one or more questions.

MAKE A TESTABLE HYPOTHESIS

The scientist tries to answer a question with an educated guess, or **hypothesis**. This should be something the scientist can test.

PLAN AN EXPERIMENT

The scientist tests the hypothesis by observing nature or by conducting an experiment.

CHOOSE EQUIPMENT AND TECHNOLOGY

In planning the experiment, the scientist must decide what equipment and technology to use.

COLLECT INFORMATION

Now the scientist is ready to conduct the experiment. The scientist carefully measures and records the results.

ANALYZE THE RESULTS

The scientist analyzes the information collected from the experiment. Scientists often organize results into a table, graph or chart.

DRAW CONCLUSIONS

The scientist thinks about what the results show. The results should relate to the hypothesis the experiment is testing.

COMMUNICATE THE RESULTS

The scientist communicates the results. The scientist describes the procedures used, so they can be repeated by other scientists.

The order of these steps may sometimes change. Results from an experiment or investigation often cause scientists to form a new hypothesis.

Let's look more closely at each of these steps. This lesson will focus on designing an experiment — the first four steps on page 43. **Lesson 6** will focus on collecting and analyzing the data, and drawing conclusions — the last four steps. For example, suppose you are interested in flying paper airplanes. How would a scientist create an experiment for studying the design of a paper airplane?

ASK A WELL-DEFINED QUESTION

Only well-defined questions can be tested by an experiment. Vague questions cannot be answered in this way. Questions for an experiment must be *specific*, *factual*, and identify *exactly* what will tested by the experiment. For example, the following question is not precise enough for a specific experiment: *What is the best paper airplane*? A scientist would wonder what "best"

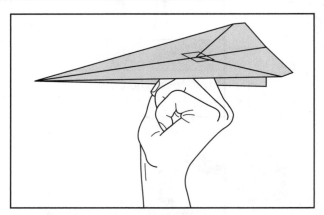

means. Does it mean the prettiest? Or does it mean the most expensive? A more well-defined question would be:

> ***Will a paper airplane fly a longer
> distance if it has a flat nose or a pointed nose?***

This question is precise. The scientist is asked to compare the effects of two specific types of "noses" on a paper airplane.

MAKE A TESTABLE HYPOTHESIS

A **hypothesis** is an educated guess that attempts to answer the question. A good hypothesis can be **tested** by an experiment. For example, a scientist may make the hypothesis that *a paper airplane will fly farther with a pointy nose than with a flat nose*. This hypothesis can be tested by an experiment. An experiment may show that the hypoth-

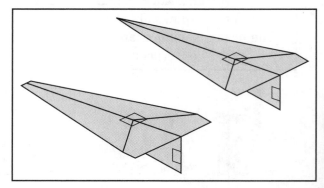

esis is either right or wrong. In science, proving a hypothesis is wrong can be just as important as proving it is right.

APPLYING WHAT YOU HAVE LEARNED

✦ Why is it just as important to prove that a hypothesis is wrong as it is to prove that it is right? _____

_____

✦ Think of an experiment you did in science class this year. What hypothesis did that experiment test? _____

_____

_____

PLAN THE EXPERIMENT

An experiment creates special conditions to test the hypothesis. It is important to plan the experiment carefully.

★ **Variables.** A **variable** is anything that can change in the experiment. In most experiments, a scientist changes one thing or variable to see what effect this has on something else. The scientist then observes the effect of this change. For example, a researcher conducting an experiment on paper airplanes can change these variables:

How do experiments build scientific knowledge?

> **What kind of paper should the airplane be made of?**

> **What is the shape of the paper airplane?**

★ In an experiment, a scientist usually changes only *one variable* at a time. For example, in the experiment involving paper airplanes, a scientist will pick one variable — such as whether the plane has a flat or pointy nose. The scientist will change **only** this variable and see how that affects something else — such as how far the airplane will fly. All other conditions should be kept exactly the same. Each plane will be the same size, be made of the same type of paper, and have the same shape except for the nose.

Keeping all other conditions the same allows the scientist to see the effects of one variable on another:

> **How do changes in one variable affect a second variable?**
>
> **Change in Variable A** → **Impact of this change on Variable B**

Often scientists have two groups in an experiment. The scientist changes something in one group (**experimental group**) but not in the other group (**control group**). Then the scientist compares the results of the two groups.

APPLYING WHAT YOU HAVE LEARNED

✦ Examine the following information:

Experimental Group	Students are given vitamin C when they have a cold.
Control Group	Students are not given anything when they have a cold.

Members of both groups record the number of days each cold lasts.

- What is being investigated? _____

- Why was a control group necessary in this experiment? _____

_____

✦ Look at the examples below. For each example, identify the question that the scientist is investigating:

Variable the scientist changes	Variable the scientist measures	What question is the scientist trying to answer?
Type of nose a paper airplane has	Distance the paper airplane can fly	
Amount of oxygen in a container	Amount of time a candle in the container will burn	
How much sand is in the soil	Amount of soil erosion in a heavy rainfall	

CHOOSE EQUIPMENT AND TECHNOLOGY

Planning an experiment is like making a recipe. First you must identify the materials, equipment and technology that are needed. Then you must list the steps to be followed to conduct the experiment.

ELEMENTS OF A GOOD EXPERIMENTAL DESIGN

★ The experiment tests the hypothesis.
★ All the variables are identified.
★ All of the required materials and equipment are listed.
★ Results can be precisely measured.
★ There should be several "trials" (times that the experiment is conducted).

A. To begin this experiment, you need: (1) two pieces of paper identical in size; (2) a 3cm piece of transparent tape; (3) a meter stick. Start by making two paper airplanes with pointy noses that are exactly the same. Here is how it is done:

Step 1: 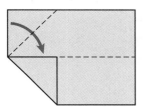 **Step 2:** 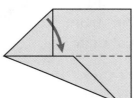 **Step 3:** 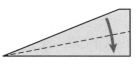 **Step 4:**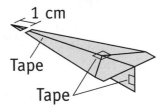

1 cm
Tape
Tape

B. Then cut off the first 1 cm piece from one of your two planes, giving it a "flat" nose. Attach the piece of paper you cut off into the fold in the center of that plane, so that its weight stays the same.

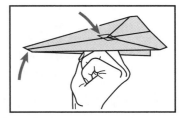

C. Now you are ready to test your airplanes. Toss each plane from an identical location using the same force.

D. Measure and record the distance each paper airplane travels until it lands. Repeat the experiment several times with each paper airplane.

Type of Plane	Distance Flown Trial #1	Distance Flown Trial #2	Distance Flown Trial #3	Distance Flown Trial #4
"Pointy-Nosed" Plane				
"Flat-Nosed" Plane				

STANDARD LABORATORY AND FIELD EQUIPMENT

The following are some of the standard types of laboratory and field equipment you should know:

Magnifiers

★ **Microscope.** An instrument that uses a series of lenses to magnify specimens placed on slides.

★ **Hand Lens.** A magnifying glass used to inspect the features of something more closely.

★ **Telescope.** An instrument that uses lenses to magnify distant objects, like stars or planets.

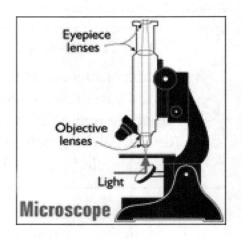

Equipment for Safety

★ **Safety Goggles.** Plastic goggles large enough to protect the eyes and face during an experiment from fine dust, splashes, mists, or sprays.

★ **Laboratory Aprons.** Bibs worn over clothing to protect clothing and the skin from splashes, spilled chemicals or biological materials.

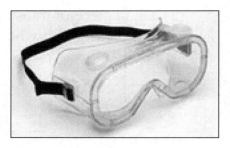

Equipment for Taking Measurements

★ **Meter Stick.** A ruler marked in centimeters (cm), used to measure length.

★ **Graduated Cylinder.** A glass cylinder marked in milliliters (mL), used to measure the volume of liquids.

★ **Thermometer.** An instrument used to measure temperatures in degrees Fahrenheit (F) or Celsius (C).

★ **Balance.** An instrument with one or two pans used to measure the mass of an item in grams (g) or kilograms (kg).

★ **Spring Scale.** A scale that measures the weight of an object by seeing how much it pulls on a steel spring attached to a dial.

★ **Timers.** Clocks, stopwatches or other devices that precisely measure the passage of time in seconds, minutes, and hours.

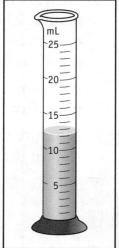

APPLYING WHAT YOU HAVE LEARNED

Equipment	How It Looks	What Is It Used For?
1. Safety goggles		
2. Microscope		
3. Hand lens		
4. Spring scale		
5. Balance		
6. Thermometer		
7. Stopwatch		
8. Telescope		
9. Graduated cylinder		
10. Meter stick		

SAFETY PRECAUTIONS

Attention to safety is essential during both field and laboratory investigations. Safety must be considered even before the experiment begins.

APPLYING WHAT YOU HAVE LEARNED

Look at the following common laboratory safety rules. Examine the list in the first column and then explain why each rule is important for safety.

Laboratory Safety Rules	Why this Rule is Important for Safety
Read all procedures before starting a laboratory investigation.	
Identify all potential hazards that could occur and take adequate safety precautions before you begin.	
Wear safety equipment when working with chemicals that could spill or splash.	
Don't begin an experiment until your teacher has given directions.	
Don't taste, smell or touch any unknown substances without directions from your teacher.	
Follow all steps, procedures and directions exactly when you conduct an experiment.	
If you are heating liquids, point them away from yourself and from other students.	
If an accident occurs, tell your teacher immediately.	
Clean up your work area and put materials away after you are done.	
Wash your hands with soap before and after all experiments.	

APPLYING WHAT YOU HAVE LEARNED

✦ How good are you at identifying some of the basic safety signs often found in a classroom laboratory? Identify each of the safety signs below:

_____   _____   _____   _____   _____

✦ Suppose you wanted to design an experiment to see the effect of ABC's fertilizer on the growth of lima bean plants. The materials for your experiment are two large flower pots with soil, lima bean seeds, a sunny garden, water, a watering can, a meter stick, and a bag of ABC's "Miracle Fertilizer."

★ What variable would you change? _____

_____

★ What conditions would you keep the same? _____

_____

★ What steps would you take to carry out your experiment?

• _____

_____

• _____

_____

• _____

_____

• _____

★ How would you take safety into account while conducting this experiment?

_____

In this lesson, you just learned how to design an experiment to test a hypothesis. In the next lesson, you will learn how to measure and analyze your results, draw conclusions, and communicate your results to others.

WHAT YOU SHOULD KNOW

◼ You should know that science asks questions about the natural world and tries to answer them by using special scientific methods.

◼ You should know that scientists use certain steps to carry out an experiment:

★ Scientists observe nature and ask a well-defined question.

★ From these observations, scientists will develop a **hypothesis** or educated guess to try to answer the question.

★ Scientists then design an experiment to test the **hypothesis**. Often the experiment changes one **variable** to see what effect this has on another **variable**. Other experimental conditions are kept the same.

★ Scientists **measure** their results. Laboratory experiments help them take exact measurements.

★ Scientists **analyze** their data and **draw conclusions**. Their conclusions should relate to the original hypothesis.

★ Scientists **communicate** their results to others.

◼ You should know that in conducting an experiment or investigation, scientists always take **safety** into account.

CHAPTER STUDY CARDS

Steps in a Scientific Investigation

★ Ask a well-defined scientific question
★ Form a testable hypothesis
★ Design an experiment to test the hypothesis
★ Select and use equipment and technology
★ Collect data
★ Analyze data
★ Form conclusions
★ Communicate results

Laboratory and Field Investigations

★ **Hypothesis.** An educated guess that tries to answer a question under investigation.

★ **Variable.** Something that can be changed or varied to find how that change affects other things in the experiment.

★ **Laboratory Equipment.** These tools include a balance, meter stick, thermometer, hand lens, and graduated cylinder.

★ **Laboratory Safety.** Safety is an important concern in all experiments and field investigations.

CHECKING YOUR UNDERSTANDING

1. Juan gives one group of 10 chickens a type of chicken feed that has no protein. He gives a second group of 10 chickens the exact same amount of chicken feed each day, but adds a small amount of protein powder. He weighs each group of chickens at the beginning of the experiment. Two months later he weighs the chickens a second time.

What question is Juan trying to answer in this experiment?

A. Will protein powder cause chickens to live longer?
B. Does protein powder cause chickens to gain weight?
C. Do chickens prefer chicken feed with protein powder?
D. Can chickens become stronger by eating protein powder?

SWK: B
G5.4

HINT

This question looks at the methods used by scientists to carry out a scientific investigation. You should understand that an experiment usually looks at the effects that changing one thing or variable has on another variable. In this question, only one variable has been changed — some chickens have protein powder added to their feed. The scientist then weighs the chickens. The experiment tests whether chickens eating protein powder will gain more weight. Therefore, the correct answer is **Choice B.**

Now try answering some additional questions on your own:

2. Sarah adds vinegar to a container of baking soda in her science class. Bubbles of carbon dioxide form. What safety precaution should Sarah take while conducting this experiment?

A. wear safety goggles
B. wash her eyes with eyewash
C. list the materials at the end of the experiment
D. leave her materials for the next class to clean up

SI: C
G3.4

3. A student wants to measure the time it takes him to run fifty meters. What instrument should the student use?

   A. a meter stick
   B. a graduated cylinder
   C. a telescope
   D. a stopwatch

   SI: A
   G4.1

   Use the information in the box below to answer questions 4 and 5.

   > *Marylou places a wooden spoon and a stainless steel spoon in a container of boiling water. She waits five minutes and then measures the temperature at the end of the handle of each spoon.*

4. Which instrument should Marylou use to measure the temperature of the spoon handles?

   A. a telescope
   B. a meter stick
   C. a thermometer
   D. a balance scale

   SI: A
   G4.1

5. In the above experiment, what question is Marylou trying to answer?

   A. Can water conduct heat?
   B. Does wood conduct heat better than steel?
   C. Does wood conduct electricity better than steel?
   D. Does heat cause either wood or steel to expand?

   SW: B
   G5.4

6. James has decided to investigate whether the number of flowers on a plant will increase if the water given to the plant is increased. James has four pots of geraniums to use in this experiment. What condition should be varied for the four plants in order to answer the question?

   A. amount of water
   B. age of the seeds
   C. temperature of the water
   D. number of hours in sunlight

   SI: C
   G5.4

7. A student sees a laboratory sign in science class. According to this symbol, what should the student do when performing an experiment in class?

   A. always wear safety goggles
   B. avoid wearing dark clothing
   C. keep away from certain foods
   D. wear a protective apron over clothing

   SI: C
   G3.4

Alice places a magnet next to a metal fork. She then records what happens. Next, she places the same magnet next to a fork made of plastic. She records what happens next.

8. Which question is Alice most likely exploring with this experiment?

    A. How do different materials react to magnets?
    B. What causes magnets to be attracted to metal objects?
    C. Does the size of the magnet affect the magnet's power?
    D. Do some objects other than magnets have magnetic power?

    SI: C
    G4.3

    SI: C
    G3.4

9. A group of fifth-graders are doing a science experiment at their work stations during science class. Which of these students is NOT practicing good laboratory safety?

    A.   B.   C.   D.

10. Almost every experiment begins with a scientist setting out to answer a particular question.

    In your **Answer Document**, identify one question a scientist might try to answer with an experiment.

    Then, identify one variable the scientist might change in an experiment designed to answer this question. (2 points)

    SI: C
    G4.3

11. When conducting a scientific experiment, a scientist follows certain safety procedures in carrying out the investigation.

    In your **Answer Document**, identify two safety procedures a scientist should follow in conducting an experiment.

    Then, explain why each procedure you selected is important for safely carrying out an experiment. (4 points)

    SI: C
    G3.4

LESSON 6

SCIENTIFIC INQUIRY: CONDUCTING AN INVESTIGATION

You just learned how scientists design experiments to test hypotheses and answer scientific questions. In this lesson, you will learn how scientists take precise measurements, analyze the data collected, draw conclusions, and communicate their results.

— MAJOR IDEAS —

A. Scientists use *rulers*, *balances*, *spring scales*, *graduated cylinders*, and *thermometers* to take measurements and collect data.

B. Scientists generally use metric units of measurement.

C. Scientists often organize their results into graphs, tables, and charts to reveal patterns in the data. Scientists interpret their data to draw conclusions.

D. Scientists communicate their results and conclusions to others. Other scientists should be able to repeat the same experiment or investigation.

E. Comparisons with other experiments are not fair unless each experiment is exactly the same. Unexpected differences often lead to different results.

COLLECTING DATA

To conduct an experiment, scientists bring their equipment and materials together in a specific order. Then they watch for results. During the experiment, scientists take precise measurements to understand what is occurring. They often measure:

Length

Temperature

Weight

Area

Volume

56

UNITS OF MEASUREMENT

Taking precise measurements is an important part of most scientific investigations. In everyday life, most Americans use the "English" system of measurement:

Length	Weight	Volume	Temperature
inch, foot, yard, mile	ounce, pound, ton	pint, cup, quart, gallon	degrees Fahrenheit (°F)

Scientists generally use the **metric system** to take measurements:

★ **Length.** A **meter** is just over 3 feet. One **inch** is equal to 2.54 centimeters.
  • 10 millimeters (mm) = 1 **centimeter** (cm)
  • 100 centimeters (cm) = 1 **meter** (m)
  • 1,000 meters (m) = 1 **kilometer** (km)

★ **Mass.** A **kilogram** is about 2.2 pounds.
  • 1,000 grams (g) = 1 **kilogram** (kg)

★ **Volume.** A **liter** is just a little more than 1 quart.
  • 1,000 milliliters (mL) = 1 **liter** (L)
  • 1 cubic centimeter (cm^3) = 1 **milliliter** (mL)

★ **Temperature.** Scientists measure temperature in degrees **Celsius**.
  • 100°C = one hundred degrees Celsius (temperature at which water boils)
  • 0°C = zero degrees Celsius (temperature water freezes at)

These prefixes show some of the relationships between metric units:

milli = one thousandth	**centi** = one hundredth	**kilo** = one thousand

APPLYING WHAT YOU HAVE LEARNED

◆ Which is longer?
  ★ 2 feet or 1 meter:_____   ★ 1 yard or 1 meter: _____

◆ Which is heavier?
  ★ 6 pounds or 3 kilograms:_____   ★ 100 grams or 1 kilogram: _____

◆ Which is warmer?
  ★ 100°F or 100°C: _____   ★ 40°F or 0°C: _____

MEASURING LENGTH

To find the **length** of something, scientists use a ruler or meter stick. To measure length, first find the 0 mark of the ruler. Then place your object starting at this mark. Look at where the object ends on your ruler. Write down the number just below that ending point. Then count the narrow lines between that number

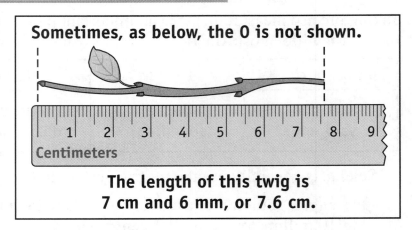

Sometimes, as below, the 0 is not shown.

The length of this twig is 7 cm and 6 mm, or 7.6 cm.

and the end of your object. If the number is in centimeters, the lines represent millimeters. You can add the centimeters and millimeters together by using a **decimal point**: [*centimeters*] • [*millimeters*]. This gives you the length in centimeters (cm).

APPLYING WHAT YOU HAVE LEARNED

✦ What is the length of this worm?

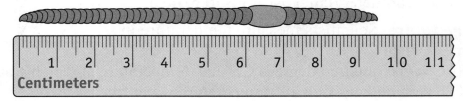

_____ cm + _____ mm. Add these two numbers together.

How many centimeters is the earthworm? _____ **cm**

✦ What is the length of the paper clip?   In inches: _____

In centimeters: _____

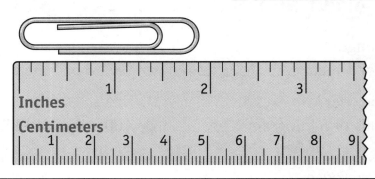

To measure the area of a rectangle, multiply its length by its width (area = L × W).

MEASURING VOLUME

Volume is how much space something takes up. To measure the volume of a liquid, scientists pour the liquid into a **graduated cylinder**. The cylinder usually has lines for each milliliter (mL), up to 100 milliliters. Often every 5 mL is labeled. The surface of the water curves up the sides of the cylinder. Measure the volume of the liquid from the **bottom** of the curve. See which line from the side of the cylinder is closest to this level.

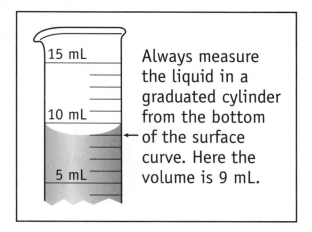

Always measure the liquid in a graduated cylinder from the bottom of the surface curve. Here the volume is 9 mL.

To measure the volume of a small solid, like a rock, add water to a graduated cylinder. Record the volume of the water. Now put the solid into the graduated cylinder. Then record the new volume. Subtract the original volume of water from the new volume. The volume of a solid object is usually recorded in cubic centimeters. Each $1 \text{ mL} = 1 \text{ cm}^3$.

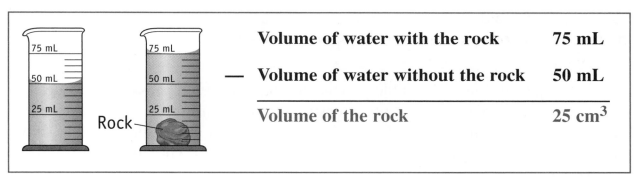

	Volume of water with the rock	75 mL
—	Volume of water without the rock	50 mL
	Volume of the rock	25 cm^3

APPLYING WHAT YOU HAVE LEARNED

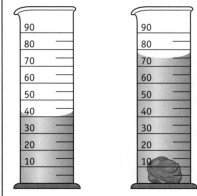

Two graduated cylinders each have the same amount of water. What is the volume of the stone in the second graduated cylinder?

_____ cm^3

WEIGHT AND MASS

Weight is how heavy something is. Weight is created by the force of gravity. It is how strongly gravity pulls the object towards Earth's center.

An object's weight is related directly to its mass. **Mass** is the amount of matter the object has. An object with greater mass is heavier than an object with less mass.

MEASURING WEIGHT

Weight is measured by a **scale**, which measures the pull of gravity on an object.

Spring Scale. Scientists often use a spring scale to measure the weight of something. The object is attached to the spring. The weight of the object stretches the spring. As it pulls, an indicator points to a number. The number on the scale shows the weight of the object.

MEASURING MASS

Scientists use a **balance** to measure mass. A balance puts a known mass on one side and the object to be measured on the other side.

Spring scale

Double-pan Balance. A double-pan balance has a bar with pans on each side. The scientist puts the object that is to be measured in one pan. In the other pan, the scientist puts known units of mass. Units of mass are added until the bar is level — both pans are the same height. The scientist then adds up all the known units to find the mass of the object.

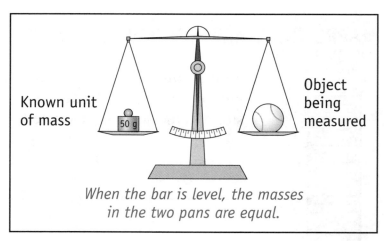

Known unit of mass

50 g

Object being measured

When the bar is level, the masses in the two pans are equal.

APPLYING WHAT YOU HAVE LEARNED

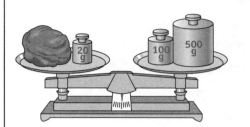

The pans in this double-pan balance are at the same height. What is the mass of the rock being measured on the left side?

_____ grams

A Triple-Beam Balance. A triple-beam balance has a single pan. Three scaled beams with "riders" are used to measure the mass of whatever is placed in the pan. The riders are moved along notches on each beam. To find the mass of the object in the pan, simply add up the numbers shown on the three riders of the triple-beam balance.

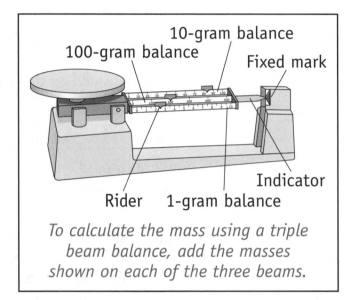

To calculate the mass using a triple beam balance, add the masses shown on each of the three beams.

APPLYING WHAT YOU HAVE LEARNED

◆ A sample is placed on a triple-beam balance. The picture below shows the riders when the scale is balanced.

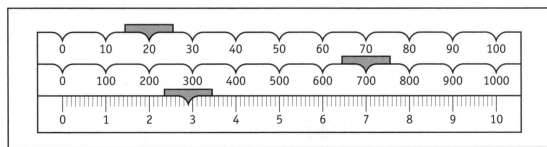

What is the mass of the sample being measured? _____

APPLYING WHAT YOU HAVE LEARNED

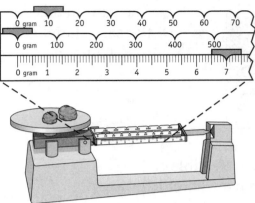

✦ What is the mass of the rocks on this triple-beam balance? _____

MEASURING TEMPERATURE

Temperature tells how hot or cold an object is. To measure temperature, scientists use a **thermometer**. There are many kinds of thermometers.

Many thermometers are glass tubes with colored liquid inside and numbers printed along the side of the tube. As the thermometer gets warmer, the liquid expands and rises inside the tube. To measure the temperature, look at the top of the liquid in the tube. Keep your eye at the same level as the liquid. Then read the number of degrees closest to that point. Scientists usually use the Celsius scale. Water freezes at 0° C and boils at 100° C.

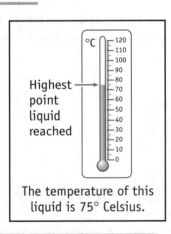

The temperature of this liquid is 75° Celsius.

APPLYING WHAT YOU HAVE LEARNED

✦ What is the temperature of each of the following thermometers?

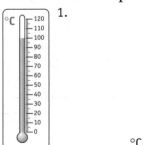

1.

_____ °C

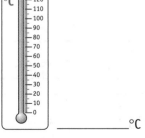

2.

_____ °C

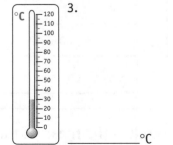

3.

_____ °C

✦ Which of these thermometers shows the boiling point of water? _____

ORGANIZING AND ANALYZING DATA

Once scientists take measurements, they record their results. Then they look to see if there are any patterns in the data. To help them find patterns, scientists often organize their results in the form of a table, graph, or chart.

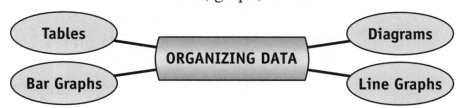

TABLES

Tables list information in columns and rows. To interpret a table, pay close attention to its headings. A scientific table often shows the relationship between two variables that are being measured. For example, look at the table below. It is based on the experiment with paper airplanes discussed in the last lesson. This table shows the relationship between the type of paper airplane and how far it flew:

TYPE OF AIRPLANE	DISTANCE AIRPLANE FLEW			
	Trial 1	Trial 2	Trial 3	Trial 4
"Flat-nosed" Airplane	4 m	5 m	4 m	3 m
"Pointy-nosed" Airplane	6 m	5 m	7 m	6 m

APPLYING WHAT YOU HAVE LEARNED

◆ What distance did the flat-nosed airplane fly on its second trial? _____

◆ What was the average distance flown by the pointy-nosed airplane on each trial? (*To find an average: add up all the numbers, then divide the total by the amount of numbers.*) _____

◆ What conclusion can you draw from this data? _____
_____
_____

BAR GRAPHS

A **bar graph** is made up of bars of different lengths. Each bar represents a quantity of something. Each bar is labeled or a key is provided to tell what each bar represents. Look at the bar graph below. It shows the average distance flown by each of the two types of paper airplanes — the pointy-nosed and flat-nosed airplane. These averages are based on the distances shown on the table on page 63.

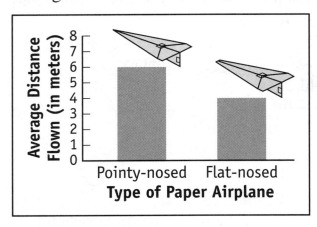

★ What was the average distance flown by the flat-nosed airplane? _____

★ Which of the airplanes was able to fly a greater distance?_____

★ How does a bar graph differ from a table? _____

_____

_____

LINE GRAPHS

A **line graph** shows a series of connected points on graph paper or a similar grid. Each point on the grid represents an amount. A line graph is usually labeled along its bottom line and left side. Examine the following line graph:

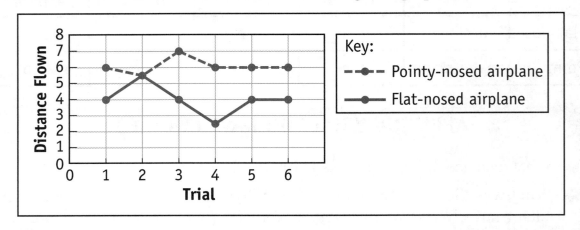

★ Based on the graph, how far did the pointy-nose airplane fly on Trial 6?

_____

★ Why do scientists often use several trials when they conduct experiments?

_____

Often a line graph is used to show how something changes over time, or how changes in one variable affect another variable. Examine the line graph on the right. It shows how the temperature of the ocean changed as scientists went deeper into the water.

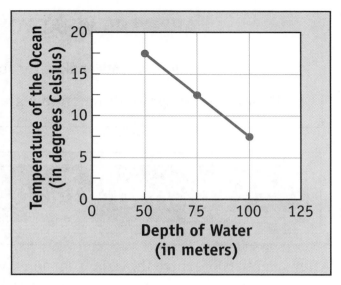

★ What is the temperature of the water at the 75 meters deep?

_____

★ How do these two variables appear to be related?

_____

_____

APPLYING WHAT YOU HAVE LEARNED

In a line graph, it is sometimes possible to guess at missing variables. Just look at the pattern in the graph and assume it continues.

◆ What do you think the temperature of the ocean is at 60 meters deep? _____

◆ What do you think the temperature of the ocean is at 125 meters deep? _____

DIAGRAMS

Diagrams can take a variety of forms. Most use pictures to show how things relate to one another. Lines or arrows often indicate relationships. For example, a group of scientists investigated prairie animals. They observed what those animals eat and recorded their observations. The diagram below summarizes their findings:

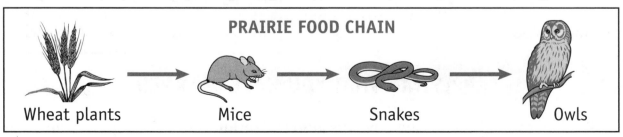

PRAIRIE FOOD CHAIN

Wheat plants → Mice → Snakes → Owls

This diagram shows that mice eat wheat plants for food. Snakes then eat the mice. The diagram shows the relationship of these prairie plants and animals.

APPLYING WHAT YOU HAVE LEARNED

◆ What prairie animal eats snakes for food? _____

◆ How does a diagram differ from a graph? _____

DRAWING CONCLUSIONS

After scientists organize the data from an experiment or investigation, they draw conclusions. The most important conclusion is about the **hypothesis** being tested.

★ Do the results of the experiment show that the hypothesis is correct or incorrect?

★ Do they suggest how the hypothesis might be changed or do they suggest a new hypothesis?

In addition, scientists also find they can sometimes make generalizations or predictions based on their results.

MAKING GENERALIZATIONS

A **generalization** is a general statement that summarizes what several pieces of data show. For example, look at the graph to the right. This is the same graph you examined on page 65. From this data, scientist might generalize that as you go deeper into the ocean, the temperature of the water becomes colder.

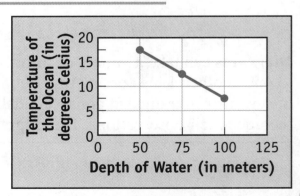

APPLYING WHAT YOU HAVE LEARNED

◆ Give examples showing how this **generalization** is true for depths from 50 to 100 meters: _____

MAKING PREDICTIONS

Scientists also make predictions from the data. A **prediction** states what will probably happen in the future. For example, as you saw, as the ocean becomes deeper, its temperature becomes colder. Based on this generalization, we can predict the ocean temperature will be colder at 125 meters than at 100 meters. We can even guess that the temperature at this depth is likely to be around 3°C.

Based on the information in the graph, what do you think the temperature of the ocean is at a depth of 65 meters? This temperature should be **less** than the temperature at 50 meters, but **greater** than the temperature at 75 meters.

APPLYING WHAT YOU HAVE LEARNED

✦ A scientist conducted an experiment growing nine bean plants in different amounts of light. The results of the experiment are shown in the line graph:

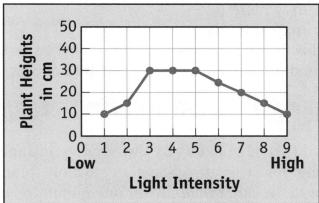

Based on the data, what conclusion can a scientist draw from this experiment?

_____

_____

COMMUNICATING RESULTS

Once scientists complete an experiment, they must communicate their results to others. There are a variety of ways in which scientific findings can be reported. Usually this takes place in the form of a ***written report***, ***article***, or an oral ***presentation***. A good presentation uses clear language and provides accurate data.

The presentation or written report also must include the exact procedures followed by the scientist so that other scientists can *repeat* the experiment and see if they obtain the same results. Scientists may also include photographs or diagrams showing what steps they performed and displaying their results.

A good scientist should communicate the results to other scientists.

It is important to keep the conditions the same when repeating an experiment. If other scientists repeat the experiment but do not keep all the conditions the same, they may reach very different results.

Differences in what is investigated, in the methods used, or in the circumstances in which the experiment occurs may produce different data. For this reason, comparisons with other experiments may not be fair when some of the conditions in each experiment are not exactly the same.

APPLYING WHAT YOU HAVE LEARNED

Scientist A placed two bean plants by the window in June and watered each plant the same amount every day. She also gave one plant fertilizer, but did not give fertilizer to the other plant. After one month the bean plant with fertilizer grew 10 cm. The bean plant without fertilizer grew only 5 cm. The scientist concluded that giving bean plants fertilizer increases their growth. Scientist B decided to repeat the experiment. In December, he placed two bean plants by the window. He watered each plant the same amount. He gave the same amount of fertilizer as Scientist A to the first plant and none to the second plant. After one month, the first plant had grown only 4 cm in height, and the second plant had only grown 2 cm in height.

✦ Why do you think Scientist B arrived at different results in this experiment?

_____

✦ Why are comparisons of the results from different experiments unfair when not all of the experimental conditions are the same? _____

_____

OBSERVATIONS AND MEASUREMENTS MADE BY OTHERS

In reviewing scientific investigations by others, scientists are always aware of the possibility of human error. A scientist may have measured something incorrectly, or simply copied an incorrect number.

It is for this reason that scientists look over data from other scientists very carefully. They are looking to see if there are any observations or measurements that do not "fit." Data that does not fit is not in the same pattern as the other results. If they find data that does not fit, scientists next try to decide whether errors were made in taking these measurements. If not, a new hypothesis may be needed to explain all of the other experimental results.

Scientists often review the data of others for possible errors.

WHAT YOU SHOULD KNOW

You should know that scientists use rulers, balances, spring scales, graduated cylinders, and thermometers to take measurements and collect data.

You should know that to measure **length**, a ruler, meter stick or centimeter ruler is used. To measure **volume**, you should use a graduated cylinder. To measure **mass**, use a double-pan or triple-beam balance. To measure **temperature**, use a thermometer.

You should know that scientists organize and analyze data by making tables, graphs, and charts. You should be able to read and interpret each of these.

You should know how to draw conclusions from data. Some conclusions from the data should relate to the hypothesis.

- To make a **generalization**, look at the data and decide what it shows.
- To make a **prediction**, guess what the data would show based on what you already know.

Scientists communicate their results and conclusions to others. Other scientists should be able to repeat the experiment.

Comparisons with other experiments are not fair if conditions in each experiment are not exactly the same. Unexpected differences in conditions will often lead to different experimental results.

CHAPTER STUDY CARDS

Measuring Data

★ **English System**
- **Length:** inches, feet, yards, miles
- **Volume:** cups, pints, quarts, gallons

★ **Metric System**
- **Length:** mm, cm, m, km.
- **Volume:** mL, cm^3
- **Mass:** grams (g) or kilograms (kg)
- **Temperature:** Degrees in Celsius (°C)

★ **Laboratory Equipment**
- **Length:** ruler, meter stick
- **Volume:** graduated cylinder
- **Weight:** spring scale
- **Mass:** double pan or triple-beam balance
- **Temperature:** thermometer

Analyzing Data

★ **Ways of Displaying Data:**
- **Table**
- **Line Graph**
- **Bar Graph**
- **Diagram**

Drawing Conclusions

★ The experiment should prove or disprove the hypothesis. The results may suggest ways to change the hypothesis for further testing.

- **Generalization.** Describes what the data shows.
- **Prediction.** States what the data would be in a new situation, based on what the data shows.

CHECKING YOUR UNDERSTANDING

1. Scientists combined two chemicals in water. The chart to the right shows the temperature of the mixture during the first five minutes after the chemicals were mixed. In the next minute, what will probably happen to the temperature?

   A. It will decrease.
   B. It will remain the same.
   C. It will increase to 200°C
   D. It will increase to 600°C

   SI: B
   G4.2

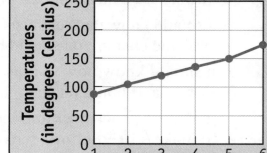

MIXING CHEMICALS

HINT

This question looks at the methods used by scientists to draw conclusions. The data shows a steady increase in the temperature of the mixture. Based on this trend, scientists are most likely to predict that the mixture will continue to increase in temperature to about 200°C. Therefore, the answer is **C**.

Now try answering some additional questions on your own:

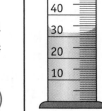

2. If five milliliters (5 mL) of vinegar are added to the water in the graduated cylinder shown on the right, what will be the total volume of the liquid?

   A. 27 mL   C. 35 mL

   B. 30 mL   D. 37 mL

   SI: B
   G4.2

3. Which race is the same length as a 1,000 meter race?

   A. a 1 kilometer race   C. a 100 kilometer race

   B. a 10 kilometer race   D. a 1,000 kilometer race

   SI: A
   G3.1

4. What units would you use to measure the mass of a soccer ball?

   A. kilometers   C. Celsius

   B. millimeters   D. grams

   SI: A
   G4.1

5. The chart below shows the time it took for bean seeds to sprout at different temperatures. Based on this data, when will seeds at 5°C most likely sprout?

   A. 5 days
   B. 8 days
   C. 13 days
   D. 18 days

   SI: B
   G4.2

Temperature (°C)	Days for Seeds to Germinate
25	5
20	7
15	9
10	11
5	?

6. Which metric measurement is closest to the height of the plant to the right based on the scale shown?

   SI: A
   G3.1

   A. 2.5 centimeters
   B. 5 centimeters
   C. 7.5 centimeters
   D. 10 centimeters

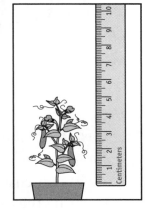

7. A scientist measures the lengths of 100 tadpoles from two different ponds. The scientist then calculates the average tadpole length in each pond. What would be the best way for the scientist to display the average lengths?

   A. bar graph   C. table

   B. line graph   D. diagram

   SI: B
   G3.5

8. The graph to the right shows the length that a plant grew over a four week period. According to the graph, how many additional centimeters did the plant grow from Week 2 to Week 4?

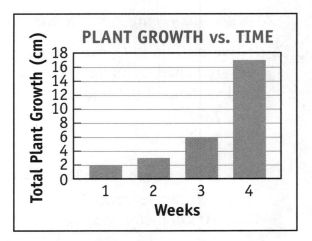

PLANT GROWTH vs. TIME

A. 6 centimeters
B. 10 centimeters
C. 14 centimeters
D. 17 centimeters

SI: B
G3.3

9. Kara did an experiment to find out what effect temperature has on the activity of yeast. Which of these steps should come last in her experiment?

A. Add 1 gram of yeast to each bowl.
B. Observe what happens in each bowl.
C. Put 120 milliliters of water in each bowl.
D. Move one bowl with yeast to a warmer location.

SI: C
G4.3

Use the graph below to answer questions 10 and 11.

A class recorded the outdoor temperature at noon on the first day of the month all through the school year. They made a graph of their results.

10. What was the temperature on the first day of March?

A. 30°C
B. 35°C
C. 40°C
D. 75°C

SI: B
G3.3

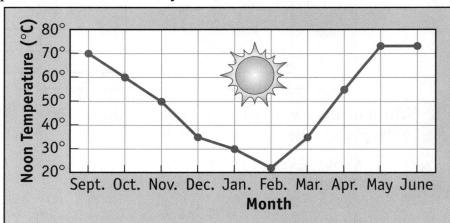

11. Based on the graph, what happened to the temperature from October to January?

A. It decreased.
B. It increased.
C. It remained the same.
D. It decreased then increased.

◆ Examine the Question
◆ Recall What You Know
◆ Apply What You Know

SI: B
G4.2

LESSON 7

SCIENCE AND TECHNOLOGY

In this lesson, you will learn about technology and its impact.

— MAJOR IDEAS —

A. **Technology** helps extend human abilities. It can have both beneficial and harmful effects.

B. Technology and science are closely related.

C. To design new technology, people first identify a problem or need, then identify possible solutions, and finally design a solution. Sometimes they revise an existing design.

D. Technology designed to solve one problem may create other problems.

WHAT IS TECHNOLOGY?

Technology is the use of tools and techniques for making and doing things. Technology is as old as humankind itself. During the Stone Age, people made tools of wood, stone, animal skins or bones to meet their needs. Early humans made arrowheads by chipping the sides of stones. Then they attached these arrowheads to sticks of wood. They used animal fibers and curved sticks to make bows. They used these simple bows and arrows to hunt animals for food or to fight against one another.

The first humans made arrows by sharpening stones.

With the introduction of farming, human technology became more complex. People learned to plant seeds to grow food. They used new tools to break up the soil, to spread the seeds, and to bring in the harvest. People could now live in one place, so they built permanent homes.

Technology helps to increase human abilities. For example, the invention of the lever made it possible for people to lift heavy objects. The invention of the wheel helped people move goods. The invention of the sailboat made it easier to move goods across lakes and seas. Technology has even helped extend the human senses. The invention of the telescope, for example, helped people to see farther than before. The telephone enabled them to listen to people thousands of miles away.

THE IMPACT OF TECHNOLOGY

The introduction of new technology has had both beneficial and harmful effects.

BENEFICIAL EFFECTS

A variety of new technologies have greatly improved people's lives.

AGRICULTURE
Some of the most important early inventions were in agriculture. Improvements in technology like the tractor and pesticides have enabled farmers to grow great quantities of food with less human effort.

Wheat harvested by a newly invented reaper.

TRANSPORTATION
Since the 1800s, transportation has been greatly improved by such technological advances as the steamboat, train, automobile, and airplane.

MANUFACTURING
Starting in the 1800s, technology helped people greatly increase their ability to manufacture goods. Using large, powerful machines powered by steam, manufacturers produced vast amounts of cloth. These same technologies were gradually applied to producing of every type of good — from clothes to cars.

COMMUNICATION
The printing press, radio, telephone, television, and the Internet have greatly improved people's ability to communicate. A person in one part of the world can now instantly communicate with someone else in almost any other part of the world.

ENTERTAINMENT

People have used technology to develop new forms of entertainment. These new forms include art, music, theater, movies, and television.

NUTRITION

Technology has enabled scientists to develop more nutritious foods, and vitamins.

HEALTH CARE

Doctors and hospitals increasingly rely on technology. Technological advances have led to X-rays, MRIs, and CAT scans, which allow doctors to see inside the body without surgery. Special machines help patients breathe during operations. Vaccines and antibiotics help patients to fight deadly diseases.

HARMFUL EFFECTS

Technology is not always beneficial. Just as a hammer can be used to smash as well as build things, technology itself can sometimes lead to harm.

THE NATURE OF WORK

Improvements in technology changed the nature of work. Most people once worked on farms or in traditional crafts. Now many of these jobs are performed by machines. Other workers face boring jobs in which they have no chance for creativity. Changes in the nature of work affect families as well as individuals. At one time, most members of the same family lived and worked together. After 1800, many began working outside the home in factories and other workplaces. As a result, family members spent much less time together.

WARFARE

Technology has greatly increased human destructiveness. In prehistoric times, people attacked each other with spears, arrows, or clubs. In the ancient world, improvements in the use of metals led to the invention of swords. Later, the development of gunpowder led to the use of cannons and guns. Each new weapon was vastly more destructive than previous ones. In 1945 the first atomic bomb was exploded. Atomic weapons have made warfare much more dangerous. These weapons threaten to destroy life on Earth.

The atomic bomb on Hiroshima killed thousands of people.

THREATS TO THE ENVIRONMENT

Technology also now threatens Earth's environment. Improvements in agriculture, nutrition and health care have led to an explosion of the world's human population. People are using larger and larger amounts of Earth's resources — air, water, and soil. Their solid and liquid wastes are polluting soil and water. People are also polluting the atmosphere by burning coal and oil to power their cars and factories, to generate electricity, and to heat their offices and homes. Pollution has led to increasing amounts of carbon dioxide in the atmosphere, bringing about "global warming."

APPLYING WHAT YOU HAVE LEARNED

Throughout history, scientists have debated whether the overall impact of technology on people's lives has been more helpful or harmful. Complete the chart below by giving two examples in each column. Then answer the question that follows.

Beneficial Effects of Technology	Harmful Effects of Technology
1._____	1._____
2._____	2._____

✦ Explain why you think the **overall impact** of technology on people's lives has either been more helpful or harmful. _____
_____
_____

SCIENCE AND TECHNOLOGY

Science is driven by a desire to understand the natural world. **Technology** is driven by a desire to meet human needs. In fact, science and technology are closely related. Scientists rely on tools developed by technology, such as telescopes, microscopes, thermometers, and glass beakers. As the level of technology improves and introduces new products — like electricity, plastics, and computers — scientists are able to use many of these new tools and materials for their own research.

Technology in turn relies on scientific discoveries. Our scientific understanding of the natural world guides the development of technology. Scientists developed the first electrical generators, x-rays, microwaves, antibiotics, jet engines, and nuclear energy. These discoveries became the basis for later inventions and technologies that we use today.

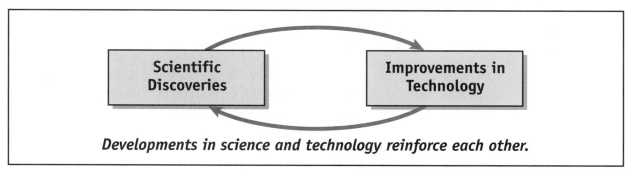

Developments in science and technology reinforce each other.

APPLYING WHAT YOU HAVE LEARNED

◆ How are science and technology alike and different? _____

_____

_____

◆ Provide an example of a new technology based on a scientific discovery:

_____

◆ Provide an example of a scientific discovery based on improvements in technology: _____

_____

THE DESIGN PROCESS

As technology improves, people's needs and wants change. People demand new and better products. This demand encourages the development of newer technologies. Technology and inventions change to meet people's wants and needs.

Scientific progress is based on the use of scientific methods of investigation. Like science, technology is based on logic and the testing of ideas. Improvements in technology depend on following the key steps of the design process:

STEPS IN TECHNOLOGICAL DESIGN

★ **Identify the Problem.** The first step is to identify a human need or problem. The new technology should help meet this need or solve the problem.

★ **Identify Possible Solutions.** Next, designers need to research the problem to look for possible solutions. They may copy solutions from elsewhere, or decide why other solutions might not work.

★ **Design a Solution.** After carefully studying possible solutions, the designer develops the best solution for this particular problem. The designer may build a model to test the solution.

★ **Evaluate and Test the Solution.** The designer finally reviews the design to see if it would really solve the problem. The designer has to keep in mind the cost of the solution. The designer also judges whether the new technology would create other problems if it were adopted. Sometimes, the design is revised to make sure it solves the problem without creating other problems. The designer may also show the proposed solution to reviewers to get their reactions and ideas.

APPLYING WHAT YOU HAVE LEARNED

Jonathan is finishing his homework on a warm spring day. The wind keeps blowing his papers around the room. He would like to hold his homework papers down so they don't blow off the table.

◆ What problem is Jonathan trying to solve? _____

_____

_____

◆ What solutions are possible for this problem? _____

_____

_____

◆ What would be the single best solution for this problem? _____

_____

_____

WHAT YOU SHOULD KNOW

You should know that technology is the use of tools and techniques to help meet human needs. It can have both beneficial and harmful effects.

You should know that science and technology are closely related

You should know that to design new technology, people first identify a problem or need, then identify possible solutions, and finally design a solution. Sometimes they revise an existing design. Technology designed to solve one problem may create other problems.

CHAPTER STUDY CARDS

The Impact of Technology
★ **Technology** is the use of tools and techniques to meet human needs.
★ **Impact of Technology.** It can have both beneficial and harmful effects.
  • **Beneficial.** Improvements in transportation, communication, agriculture, manufacturing, nutrition, health care, and entertainment.
  • **Harmful.** Some jobs are monotonous; new weapons are more destructive; the environment is threatened.

Science and Technology
★ **Science.** Desire to understand the world.
★ **Technology.** Desire to meet human needs.
★ **Science and Technology.** Both reinforce each other.

The Design Process
★ **Identify a Problem.**
★ **Identify Possible Solutions.**
★ **Design a Solution.**
★ **Evaluate and Revise the Solution.**

CHECKING YOUR UNDERSTANDING

1. In what way is the process of technological design similar to a scientific investigation?
   A. Technological designers test and revise their ideas.
   B. Technological designs must be capable of mass production.
   C. Technological designs help people understand the natural world.
   D. Technological designers change one variable to see the effect on another.

S&T: B
G4.3

HINT

This question asks you to compare the processes of technological design and scientific investigation. You have to identify something they have in common. Recall what you know about each process. Technological designers identify a need, and create and revise their designs. Scientists ask questions, develop a hypothesis, and test the hypothesis in an experiment. Now apply what you know to the question. **Choice A** correctly identifies an important similarity between the two processes.

Now try answering some additional questions on your own:

2. Which of the following has been a negative effect of technology?
   A. Farmers are able to grow more food.
   B. Pollutants have led to global warming.
   C. People enjoy music on the radio and CDs.
   D. People can easily communicate around the globe.

   S&T: A
   G3.2

3. Which is a benefit from the invention of the electric dishwasher?
   A. It saves people time and effort.
   B. It uses hot water to wash dishes.
   C. It uses detergent that can pollute the environment.
   D. It uses more electricity than washing dishes by hand.

   S&T: A
   G4.1

4. People often have difficulty holding water bottles while running. A manufacturer decides to develop a new product to help solve this problem. What step should the manufacturer take next?
   A. identify the problem
   B. decide on the solution
   C. examine many possible solutions
   D. evaluate and revise the solution

   S&T: B
   G4.3

5. Many forms of technology have both beneficial and harmful effects.

   In your **Answer Document**, identify a recent technology.

   Then, explain one beneficial or harmful effect of this technology. (2 points)

   S&T: A
   G3.2

6. The fields of science and technology are closely related. Many modern technological wonders used in everyday life are based on scientific work.

   In your **Answer Document**, identify one way in which science has influenced technology.

   Then, identify one way in which technology has influenced science. (2 points)

   S&T: A
   G4.2

CONCEPT MAP FOR THINKING LIKE A SCIENTIST

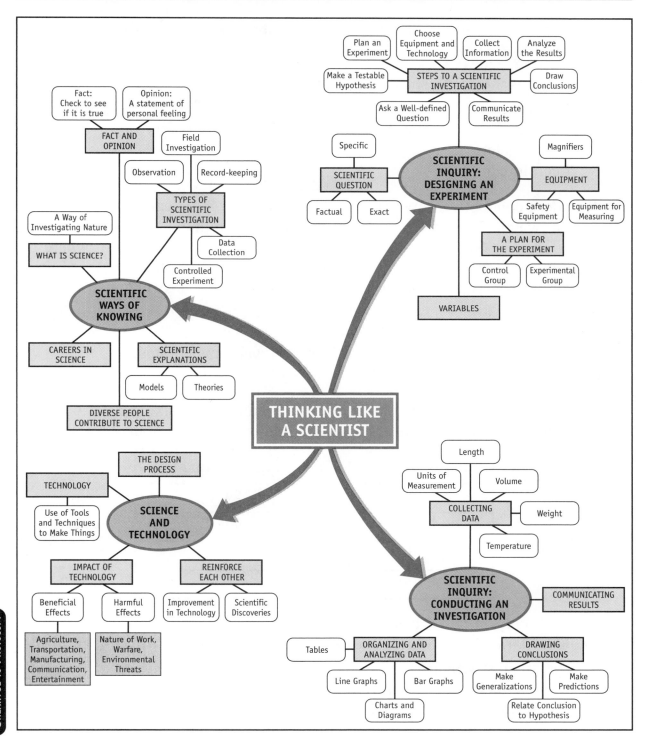

TESTING YOUR UNDERSTANDING

> *At the end of each content unit you will find a **Testing Your Understanding**. The purpose of this section is to provide practice questions touching on the entire unit.*

1. How long is the screw shown in the picture?

    A. 4 centimeters

    B. 5 centimeters

    C. 6 centimeters

    D. 7 centimeters

(SI: A G4.1)

2. A student conducts an investigation by filling two identical beakers with 200 mL of water. The student places one beaker in a dark room where the temperature is 68°F. She places the second beaker in a dark room where the temperature is 50°F. Six hours later, the student measures the amount of water in each beaker. Which question does this investigation answer?

    A. Does light affect how fast water evaporates?

    B. Does the time of day affect how fast water evaporates?

    C. Does a container's shape affect how fast water evaporates?

    D. Does surrounding temperature affect how fast water evaporates?

(SW: B G5.4)

3. The table below shows the movement of a snail during a six hour period. If the snail moves at the same pace, how far would it travel in 8 hours?

    A. 40 cm

    B. 120 cm

    C. 160 cm

    D. 180 cm

(SI: B G4.2)

Time Traveled	Total Distance Traveled
2 hours	40 cm
4 hours	80 cm
6 hours	120 cm
8 hours	?

4. What conclusion could best be drawn from a study of the work of Aristotle, Galileo, Marie Curie, Shirley Jackson, and Chien-Shiung Wu?

    A. The greatest scientists have come from China.

    B. Women made greater contributions to science than men.

    C. Most important scientific discoveries have already been made.

    D. People from many different cultures have contributed to science.

(SW: D G3.3)

5. Which statement is a testable hypothesis?

    A. Apples are good for you.
    B. Soup tastes better when eaten with salt.
    C. Sound travels faster through water than air.
    D. There are many good designs for paper airplanes.

SI: C
G4.3

6. Which statement is an opinion?

    A. The universe may collapse someday.
    B. The moon has both a day and a night.
    C. Earth orbits the sun once every 365 days.
    D. Stars are made up of gases that are super-heated.

SW: A
G4.1

The following table shows data collected during an experiment involving temperatures recorded by students in Toledo from Monday to Friday.

Day	Monday	Tuesday	Wednesday	Thursday	Friday
Temperature	72° F	77° F	78° F	82° F	70° F

7. Which thermometer correctly shows the temperature recorded on Friday?

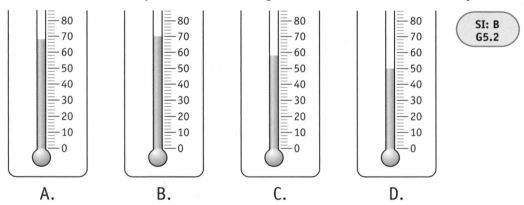

    A.   B.   C.   D.

SI: B
G5.2

8. A scientist writes an article for a scientific journal describing an experiment he has conducted. Why is it important that he clearly describe all of the steps he followed in conducting the experiment?

    A. Scientists should be able to repeat the experiment.
    B. Comparisons between experiments are often unfair.
    C. Scientists often make predictions based on patterns in the data.
    D. His conclusions should either prove or disprove his hypothesis.

SW: C
G4.4

9. Which piece of laboratory equipment should be used to find the volume of a small stone?

    A. graduated cylinder   C. meter stick
    B. thermometer   D. spring scale

SI: A
G4.1

CHECKLIST OF SCIENCE BENCHMARKS

*At the end of each content unit you will find a **Checklist of Science Benchmarks** like the one below. The purpose of these checklists is to help you monitor your understanding of the major science benchmarks examined in each unit before moving on to the next one.*

Directions. Now that you have completed this unit, place a check (✔) next to those benchmarks you understand. If you are having trouble recalling information about any of these benchmarks, review the lesson indicated in the brackets.

SCIENTIFIC INQUIRY

☐ You should be able to use appropriate instruments safely to observe, measure and collect data when conducting a scientific investigation. [**Lessons 5 and 6**]

☐ You should be able to organize and evaluate observations, measurements and other data to formulate inferences and conclusions. [**Lesson 6**]

☐ You should be able to develop, design and safely conduct scientific investigations and communicate the results. [**Lessons 5 and 6**]

SCIENTIFIC WAYS OF KNOWING

☐ You should be able to distinguish between fact and opinion and explain how ideas and conclusions change as new knowledge is gained. [**Lesson 4**]

☐ You should be able to describe different types of investigations and use results and data from investigations to provide the evidence to support explanations and conclusions. [**Lesson 4**]

☐ You should be able to explain the importance of keeping records of observations and investigations that are accurate and understandable. [**Lessons 4 and 6**]

☐ You should be able to explain that men and women of diverse countries and cultures participate in careers in all fields of science. [**Lesson 4**]

SCIENCE AND TECHNOLOGY

☐ You should be able to describe how technology affects human life. [**Lesson 7**]

☐ You should be able to describe and illustrate the design process. [**Lesson 7**]

UNIT 3

LIFE SCIENCES

In this unit, you will review what you need to know for the **Grade 5 Science Achievement Test** about the life sciences — the study of living things.

Flamingos interact with their environment.

You will learn how living things meet their needs, and how they are classified. You will also learn how plants and animals differ, and how they interact with each other and their environment in ecosystems.

LESSON 8: PLANTS

In this lesson, you will learn how plants differ from animals. You will learn how plants are classified and how plants meet their needs. You will also learn about the life cycle of plants.

LESSON 9: ANIMALS

In this lesson, you will learn about animals. You will learn how animals are classified and how they meet their needs. You will also learn about the life cycle of animals.

LESSON 10: ECOSYSTEMS

This lesson looks at how different types of plants and animals live together in ecosystems. In this lesson, you will learn how ecosystem changes affect living things, and how living things can affect their ecosystem. You will also learn how fossils provide evidence about plants and animals that once lived long ago.

LESSON 8

PLANTS: THEIR CHARACTERISTICS, TYPES AND LIFE CYCLES

In this lesson, you will learn how plants and animals differ. You will also learn how different plant structures help plants to meet their needs. You will see how scientists classify different plants. Finally, you will learn about the life cycles of plants.

— MAJOR IDEAS —

A. Plants can make their own food by using the energy from sunlight. However, plants cannot move from one place to another.

B. Plants have specific structures that help them to survive and reproduce. These include roots, stems, leaves, seeds and spores.

C. Scientists classify plants into different types, based on their characteristics.

D. All plants have their own life cycles.

WHAT IS A PLANT?

You are already familiar with plants and animals: they are all around you. But can you explain what makes plants and animals so different? What sets plants apart?

Like all living things, plants are made of tiny **cells**. Plant cells have some unique characteristics. They contain a green chemical called chlorophyll. **Chlorophyll** helps plants turn energy from the sun into food. Because plants can make their own food, they are

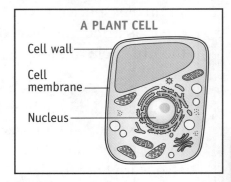

A PLANT CELL

Cell wall

Cell membrane

Nucleus

called **producers**. When scientists look at plants under a microscope, they also see that plant cells are surrounded by stiff cell walls. These walls help plants stand. However, because of their rigid cell walls, plants cannot move from place to place.

APPLYING WHAT YOU HAVE LEARNED

✦ Describe two things that make plants unique:

• Food: _____

• Movement: _____

HOW PLANTS MEET THEIR NEEDS

Like all living things, plants need to survive, grow and reproduce. All plants have three basic parts: *roots*, *stems*, and *leaves*. Each of these plant structures serves a specific function in helping the plant meet its needs.

ROOTS

Roots are the part of the plant found below the ground. They hold the plant in the ground, and they absorb water and minerals from the soil.

Plant roots absorb water.

STEMS

Stems form the main body of the plant. They support its leaves and bring them into the sunlight. In some plants, stems also hold flowers. Stems bring water and minerals to the leaves. They also carry food from the leaves to the roots. Stems can be as narrow as the stem of a flower, or as wide as the trunk of a tree.

LEAVES

Leaves make food from the sun's energy. This process is called **photosynthesis**. Tiny pores in the bottom of the leaves absorb carbon dioxide from the air and give out oxygen. In the leaves, sunlight is mixed with carbon dioxide and water to produce a type of sugar. The plant uses this sugar as food. The process of photosynthesis also produces oxygen.

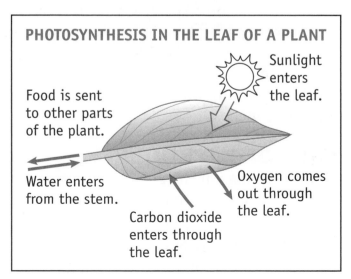

PHOTOSYNTHESIS IN THE LEAF OF A PLANT

Sunlight enters the leaf.

Food is sent to other parts of the plant.

Water enters from the stem.

Carbon dioxide enters through the leaf.

Oxygen comes out through the leaf.

APPLYING WHAT YOU HAVE LEARNED

◆ How do plants get food? _____

_____

◆ How do plants stay in the same place? _____

◆ How do plants get water? _____

Special Plant Structures. Plants often have special structures to survive. For example, some branches or leaves have sharp spines or thorns to protect a plant from being eaten. Climbing plants have long, thin stems that wrap around trees. A cactus has a thick stem to hold water and sharp spines to produce shade.

CLASSIFYING TYPES OF PLANTS

If you look at a garden, you will notice that not all plants are the same. In fact, scientists have found that there are thousands of different kinds of plants. Scientists **classify** these different kinds of plants into groups. The members of each group share common characteristics, such as stem type, leaf type or having seeds and flowers.

STEM TYPES

The most basic division of plants is based on their type of stem. Most plants have tube-like structures in their stems to carry water and food to different parts of the plant. You can see these tubes easily in a celery stalk. More primitive plants, like mosses, do not have these tube-like structures. As a result, mosses cannot grow any taller than a few inches.

Moss grows close to the ground.

WAYS OF REPRODUCING

All living things must **reproduce** to survive. Scientists divide some plants into different groups based on the ways in which they reproduce.

★ **Plants Without Seeds.** Some plants, like ferns, reproduce without seeds. Instead, they have spores. A **spore** is a cell that comes from only one parent, but can grow into a new plant.

★ **Plants With Seeds.** Most common plants have seeds. A **seed** has a tiny plant and food for it to grow. Many plants with seeds also have flowers. These are known as **flowering plants**. Flowering plants often form fruit. Animals may eat the fruit and carry the seeds inside the fruit to a new place. Some plants with seeds, like pine trees, produce cones rather than flowers or fruit.

PARTS OF A SEED

Embryo

Root   Stem   Leaf

Seed coat

LEAF TYPES

Trees can be divided into different types based on their leaves. Many trees change color and drop their leaves in autumn. These trees grow new leaves in spring. In contrast, evergreen trees have narrow leaves known as "needles." These trees do not change color or lose their needles in the winter.

APPLYING WHAT YOU HAVE LEARNED

Scientists are able to classify plants into different groups based on the characteristics of their stems, leaves, and ways of reproducing. Name two types of plants and describe the characteristics of each type. If you cannot name two plants, use an encyclopedia or the Internet to find this information.

Name of Plant	Description of Its Stem, Leaves, and Way of Reproducing

THE LIFE CYCLES OF PLANTS

Do you look the same as you did five years ago? Of course not! And you will not look the same twenty years from now. As you live, you change. As living things age, they go through changes. All living things go through steps known as **life cycles**. All living things begin life, grow, mature, reproduce, and eventually die.

Like other living things, plants go through life cycles. Here is the life cycle of a typical flowering plant:

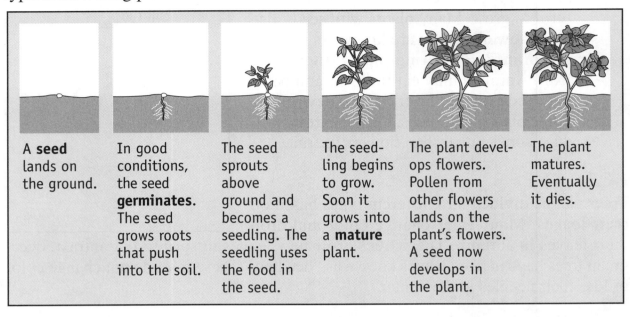

| A **seed** lands on the ground. | In good conditions, the seed **germinates**. The seed grows roots that push into the soil. | The seed sprouts above ground and becomes a seedling. The seedling uses the food in the seed. | The seedling begins to grow. Soon it grows into a **mature** plant. | The plant develops flowers. Pollen from other flowers lands on the plant's flowers. A seed now develops in the plant. | The plant matures. Eventually it dies. |

The key steps in this life cycle are *germination*, *maturity*, *reproduction* and *death*.

SPREADING SEEDS

Plants cannot move around like animals can. But seeds need to be moved away from their parent plant if they are to survive. Seeds that grow too close to a parent plant will compete with the parent plant for sunlight and water. Many seeds are blown by the wind. Some plants depend on animals to carry their pollen or spread their seeds. Many plants grow their seeds in fruits that are eaten by animals. When the seed is passed out or discarded by the animal, it grows in another place.

APPLYING WHAT YOU HAVE LEARNED

Carefully examine the illustration to the right. Number the different plants in the order of their development in the plant life cycle.

★ Step 1: _____

★ Step 2: _____

★ Step 3: _____

★ Step 4: _____

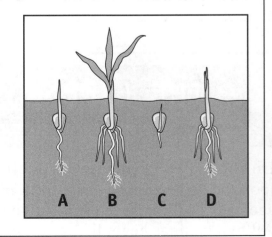

WHAT YOU SHOULD KNOW

☐ You should know that plants can make their own food with energy from sunlight (**photosynthesis**). Plants cannot move from one place to another.

☐ You should know that plants have specific structures — **roots**, **stems**, and **leaves** — which help them to grow, survive and reproduce.

☐ You should know that scientists **classify** plants into different types based on their characteristics, such as their type of stem or whether they have flowers.

☐ You should know that plants have their own **life cycles**.

CHAPTER STUDY CARDS

Plant Characteristics

★ **Photosynthesis.** Plants make their own food from sunlight through a process called photosynthesis. Plants use carbon dioxide and produce oxygen. Because plants make their own food, they are called **producers**.

★ **Movement.** Plant cells are surrounded by stiff cell walls. These rigid cell walls prevent plants from moving from place to place.

★ **Plant Classification.** Scientists classify plants based on:

• **Stem Type.** Whether the stem has tubes.
• **Seeds.** If the plant has flowers or seeds.
• **Leaves.** Whether the plant loses its leaves.

Main Parts of the Plant

★ **Roots:** anchor the plant in the ground, and absorb water and minerals.
★ **Leaves:** Make food from sun's energy.
★ **Stems:** Support leaves and flowers.

Typical Plant Life Cycle

★ A **seed** lands on the ground.
★ Seed **germinates** (sprouts) as its roots push into the soil
★ **Seedling** is formed. The seedling grows into a **mature plant** and often develops flowers.
★ **Reproduction** occurs when flowers are pollinated from other plants.
★ Plant eventually **dies**.

CHECKING YOUR UNDERSTANDING

1. Which identifies a function of this plant's roots?

   A. to make seeds
   B. to store oxygen
   C. to absorb water
   D. to produce pollen

LS: B
G4.2

> HINT
>
> To answer this question correctly, you need to recall the parts of a plant and the function of each part. The roots hold the plant firmly in the soil. In addition, they absorb water and minerals from the soil. Since these are the main functions of the roots, **Choices A**, **B**, and **D** are incorrect. The correct answer is **Choice C**.

Now try answering some additional questions on your own:

2. An oak tree can make its own food because its leaves use energy from

    A. water
    B. oxygen

    C. sunlight
    D. minerals

    > LS: B
    > G5.1

3. The diagram to the right shows a bean seed in a container of soil. The seed has started to germinate. The first part to come out is labeled **X** in the illustration.

    What conclusion can be drawn about this seed?

    A. It is ready to reproduce.
    B. It does not require water.
    C. It will not grow into a mature plant.
    D. It is in the process of becoming a seedling.

    > LS: A
    > G4.1

4. Juan and Maria want to grow carrots in their garden. What can they bury in the soil to succeed?

    A. carrot stems
    B. carrot seeds
    C. carrot leaves
    D. carrot flower petals

    > LS: B
    > G4.2

5. Dwayne decides to investigate seed germination. He plants several seeds in his garden. What will happen to these seeds when they germinate?

    A. They will grow seeds.
    B. They will attract bees.
    C. They will start to spout.
    D. They will conduct photosynthesis.

    > ◆ **Examine the Question**
    > ◆ **Recall What You Know**
    > ◆ **Apply What You Know**

    > LS: A
    > G4.1

6. In what important way do some plants in the forest depend on animals?

    A. Animals provide the plants with sunlight.
    B. Animals secure the plant roots in the soil.
    C. Animals carry the plant's pollen or spread their seeds.
    D. Animals make sure the plants will receive enough water.

    > LS: A
    > G4.5

7. Many plants produce seeds. Often these seeds are spread by the wind. Which of the seed types below is likely to be spread farthest by the wind?

| A | B | C | D |

LS: A
G4.5

8. A student plants grass seeds in a small pot of soil. What must the seeds have before they will be able to sprout?

   A. new rocks added to the soil
   B. temperatures above 80°F each day
   C. enough water to keep the soil moist
   D. sunlight for at least 6 hours each day

   ◆ Examine the Question
   ◆ Recall What You Know
   ◆ Apply What You Know

   LS: A
   G4.1

9. A student is making a diagram of a plant for science class to illustrate photosynthesis. In addition to leaves, which plant part would be the most important to include in the diagram?

   A. the bark because it protects the stem
   B. the roots because they bring in water
   C. the flowers because they attract insects
   D. the seeds because they help it reproduce

   LS: B
   G4.2

10. What role does photosynthesis have in a plant?

   A. It helps the plant to grow flowers.
   B. It provides the plant with the ability to move.
   C. It provides the reproductive organs of a plant.
   D. It is the main source of food energy for the plant.

   LS: B
   G5.1

11. Scientists use various characteristics to classify plants into different groups.

   In your **Answer Document**, identify two characteristics that scientists use to classify plants. (2 points)

   LS: B
   G4.3

12. Although there is a wide variety of plants, most plants have three main parts.

   In your **Answer Document**, identify two main parts of a plant that help it to grow, survive, or reproduce.

   Then, explain how each part helps the plant to survive. (4 points)

   LS: B
   G4.2

13. A typical flowering plant goes through four main stages in its life cycle.

   In your **Answer Document**, make a diagram showing the four stages in the life cycle of a typical flowering plant. (4 points)

   LS: A
   G4.1

LESSON 9

ANIMALS: THEIR CHARACTERISTICS, TYPES AND LIFE CYCLES

In **Lesson 8**, you learned about plants. In this lesson, you will learn about animals. You will learn how animals meet their basic needs, how they are classified, and how they pass through their own life cycles as they grow, mature, reproduce, age and die.

— MAJOR IDEAS —

A. Unlike plants, animals cannot make their own food. In order to survive, they must eat plants or other animals.

B. Animals can move from one place to another. They use their senses to guide their movement.

C. Animals have specific structures that help them to survive and reproduce.

D. Scientists classify animals into different types, based on their characteristics.

E. Animals have their own life cycles.

WHAT IS AN ANIMAL?

What makes each of these different kinds of living things, or **organisms**, an animal?

_____

_____

What All Animals Have in Common. All animals have two things in common. First, they cannot make their own food. In order to obtain energy, they must eat plants or other animals. Because they depend on eating other living things to survive, animals are called **consumers**. Second, unlike plants, animals can move from place to place. This helps them to find food and escape from enemies.

HOW ANIMALS MEET THEIR NEEDS

Like plants, animals have special structures that help them to survive.

SENSE ORGANS
Animals have senses like touch, smell, hearing and sight. These senses help animals to know when they are in danger and where they can find food.

MEANS OF MOVEMENT
All animals move. This allows them to find food and to hide or escape from their enemies. Animals have different structures to help them move. Fish, for example, have **fins** and **tails**, which they can move to swim through the water. Birds and many insects have **wings** that allow them to fly. Insects, birds, reptiles, and mammals have **legs**, which help them to walk on land.

BREATHING
Special structures help animals obtain oxygen to burn food. Land-based animals breathe oxygen from the air through their mouths and noses. Oxygen enters the blood in the **lungs**. Fish absorb oxygen from the water through their **gills**.

DRINKING AND EATING
Animals also must have food and water to survive. Special structures, such as mouths and stomachs, allow them to drink, eat and absorb nutrients and water.

BODY COVERINGS
Animals are covered by skin and scales, hair or feathers. The skin holds the body together, provides warmth, and protects animals from outside conditions.

SPECIAL CHARACTERISTICS

Each type of animal also has special characteristics that help it survive. For example:

★ **Giraffe.** A giraffe's long neck and keen eyesight allow it to see for many miles. Their long necks help them eat tree leaves too high for most other animals. Scientists believe their skin pattern makes them look like tall trees, helping them hide from animals that eat them.

★ **Alligators.** These animals have long tails, long snouts, and smooth skin. These characteristics aid the alligator in swimming. Alligators have small legs so they can walk on land. They are equally at home on land or in the water, which is perfect for the swamplands they inhabit.

★ **Camel.** A camel is a large mammal with one or two humps of body fat. Camels can store water in their blood and can live without water for two weeks. They can exist without food for a month. Camels are able to withstand changes in body temperature. Their thick coat of camel hair is able to reflect sunlight.

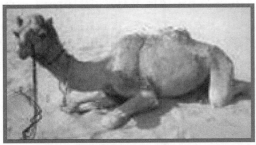

Special Animal Behavior. Many animals also have special ways of acting that help them to survive. For example, bears **hibernate** (*sleep*) in winter so they will not need more food. Many birds fly south in winter to warmer places.

APPLYING WHAT YOU HAVE LEARNED

Every animal has special characteristics that help it to survive. Choose two animals and explain how their characteristics help them to survive.

◆ _____

◆ _____

_____

DIFFERENT TYPES OF ANIMALS

Just as scientists classify plants, they also classify animals into different types.

BODY STRUCTURE

If you lift your head, you can feel a series of bones at the back of your neck. This is your **backbone**, or spine. Although you have a backbone, not all animals have one. Scientists use this as one of the most important ways to classify animals.

★ **Animals without Backbones.** These include spiders, ants, snails, earthworms, shrimp, lobsters, and jellyfish. All insects lack backbones.

★ **Animals with Backbones.** Animals with backbones are further divided into five main groups: *fish*, *amphibians*, *birds*, *reptiles* and *mammals*. You should know the chief characteristics of each of these groups.

Type	Lives on Land or Water	Body Covering	Limbs	Reproduction	Examples
Fish	Live in saltwater or freshwater. Cannot survive outside of water.	Covered with scales.	Have no legs: have fins and tails.	Usually lay eggs.	Salmon, bass, trout, tuna
Amphibians	Amphibians, like frogs, live in the water in childhood and outside water as adults.	Smooth skin without scales.	Have four legs.	Usually lay eggs.	Frogs, turtles
Reptiles	Generally live on land and have lungs for breathing.	Most have waterproof scales.	Most have four legs.	Usually lay eggs.	Alligators, snakes, lizards
Birds	Generally live on land and have lungs.	Bodies are covered with feathers.	Front legs are wings. Most are able to fly.	Lay eggs.	Chickens, eagles, ducks
Mammals	Generally live on land and have lungs. Whales and dolphins live in the water.	Most are warm blooded, and are covered with fur or hair.	Most have four legs.	Most mammals give birth to live young. All feed their young milk.	Zebras, monkeys, cats, dogs, humans

APPLYING WHAT YOU HAVE LEARNED

Check (✔) if each animal has a backbone or not. Classify those with backbones as a *fish*, *amphibian*, *reptile*, *bird* or *mammal*. If you are not sure of what kind of animal it is, you should consult an encyclopedia or the Internet.

Animal	Back/No Backbone		Type of Animal
Shark	☐ Backbone	☐ No Backbone	
Salamander	☐ Backbone	☐ No Backbone	
Mole	☐ Backbone	☐ No Backbone	
Ostrich	☐ Backbone	☐ No Backbone	
Butterfly	☐ Backbone	☐ No Backbone	

THE LIFE CYCLES OF ANIMALS

Like plants, animals go through life cycles. They are born, mature, reproduce, age and die. Reptiles, fish, insects, and birds are usually born from eggs. Almost all mammals have live births.

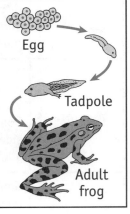

Metamorphosis. Amphibians and insects go through special stages and usually change from one form to a completely different form after birth. This is known as **metamorphosis**. For example, a frog begins as an **egg**. Eventually, the egg will hatch and form a **tadpole**. After several weeks, the tadpole will develop tiny legs and arms. Eventually, the tadpole grows into a frog.

Butterflies provide an example of **insect metamorphosis**. The cycle starts when an adult butterfly lays an **egg**. Out of the egg comes a **caterpillar** (*or larva*). The caterpillar wraps itself up in a **chrysalis** (*or cocoon*). After a period of time, an adult **butterfly** emerges from the chrysalis.

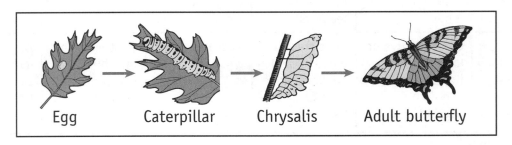

Egg → Caterpillar → Chrysalis → Adult butterfly

APPLYING WHAT YOU HAVE LEARNED

✦ Describe each of the following stages in the life cycle of a frog:

Egg		
Tadpole		
Adult Frog		

WHAT YOU SHOULD KNOW

- You should know that animals cannot produce food. They must eat plants or other animals.
- You should know that animals can move from one place to another. They use their different senses to guide their movements.
- You should know that scientists classify animals based on their characteristics, such as body coverings and structures.
- You should know that animals go through their own life cycles.

CHAPTER STUDY CARDS

Animals

★ Animals cannot make their own food. They must eat plants or other animals.
★ Unlike plants, animals move from place to place to meet their needs.
★ Animals use their senses, means of movement, structures for breathing and eating, and body coverings to survive. An animal's structures have specific survival functions.
★ **Life Cycles.** Animals go through life cycles. In **metamorphosis**, an insect or amphibian changes its form as it grows.

Animal Classification

Scientists classify animals into different types based on characteristics like body structure.

★ **Without Backbones.** Animals without backbones include jellyfish, insects, lobsters, and earthworms.
★ **With Backbones.** Animals that have backbones include fish, amphibians, reptiles, birds, and mammals. Each of these types has its own special characteristics: for example, birds have feathers.

CHECKING YOUR UNDERSTANDING

Examine the following illustrations:

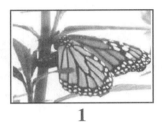

| 1 | 2 | 3 | 4 |

1. Which is the correct order of development for this butterfly?

A. $4 \rightarrow 2 \rightarrow 1 \rightarrow 3$ C. $4 \rightarrow 3 \rightarrow 2 \rightarrow 1$

B. $1 \rightarrow 4 \rightarrow 2 \rightarrow 3$ D. $3 \rightarrow 2 \rightarrow 4 \rightarrow 1$

**LS: A
G3.1**

HINT

To answer this question correctly, you must know the life cycle of a butterfly. A butterfly is an insect that goes through a metamorphosis in its lifetime. When its egg hatches, a caterpillar comes out. The caterpillar later forms a chrysalis. Out of the chrysalis comes a beautiful butterfly. If the pictures were arranged in the order of their development, the order would be: **3 – 2 – 4 – 1.** The answer is **Choice D**.

Now try answering some additional questions on your own:

2. Owls eat mice and other small animals to survive. Owls fly and hunt at night in the dark. Which structure helps owls hunt for their next meal?

A. white eggs
B. large eyes
C. strong voice
D. feathers around eyes

♦ **Examine the Question**
♦ **Recall What You Know**
♦ **Apply What You Know**

**LS: B
G3.2**

3. An anteater is an animal that survives mainly by finding and eating ants. Its main method of hunting is by sticking its long snout in the ground to find ants. Which of the anteater's senses is most useful in seeking out food?

A. its ability to see
B. its sharp hearing
C its sense of smell
D. its sense of touch

**LS: B
G3.2**

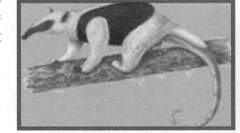

An anteater

4. Which of the following is most like the pair to the right?

   LS: A
   G3.1

   A. egg — seed
   B. branch — tree
   C. worm — snake
   D. caterpillar – butterfly

TADPOLE	FROG

5. Fish that swim in the ocean, a pet dog, and an owl are alike in important ways. What is one of the ways all three of these animals are alike?

   A. They all have legs.
   B. They all have hair.
   C. They all have eyes.
   D. They all have lungs.

   LS: B
   G3.3

6. What does every animal need in order to survive?

   A. food, water, and air
   B. eyes, nose, and ears
   C. roots, leaves, and stems
   D. light, soil, and nutrients

   ◆ **Examine the Question**
   ◆ **Recall What You Know**
   ◆ **Apply What You Know**

   LS: B
   G3.1

7. The pictures below show the main stages in the life cycle of a fly. Which group shows the life cycle of the fly in the right order?

   LS: A
   G3.1

8. Which body structure of an eagle helps it to find food by diving down in the air?

   A. its beak
   B. its neck
   C. its claws
   D. its wings

   LS: B
   G3.2

9. How would being able to change their color help some animals to survive?

   A. It would help these animals to find food.
   B. It would allow these animals to control their speed.
   C. It would be more difficult for other animals to find them.
   D. It would make it easier for babies to recognize their mothers.

   LS: B
   G3.1

10. How are animals different from plants?

    A. Animals need water and air.
    B. Animals must have hair or fur.
    C. Animals cannot make their own food.
    D. Animals have characteristics to survive their environment.

    LS: B
    G5.2

Look at the chart below showing two groups of animals:

Group A	Group B
Earthworms	Eagles
Jellyfish	Goldfish
Bumble bees	Cats

11. What characteristic has been used to classify these animals into groups?

    A. whether they can fly
    B. whether they need food
    C. whether they have a backbone
    D. whether they live on land or in the water

    LS: B
    G3.3

12. In what way are plants and animals similar?

    A. They both are mammals.
    B. They both are able to move around.
    C. They both need water to survive.
    D. They both are able to make their own food.

    LS: C
    G5.4

13. Diversity in the world exists not only among people but also among different kinds of animals.

    In your **Answer Document**, identify two characteristics that scientists use to classify animals. (2 points)

    LS: B
    G3.3

14. Animals meet their needs for survival in a number of ways.

    In your **Answer Document**, identify two structures that animals use to help them to survive.

    Then, explain how each structure functions to help them survive. (4 points)

    LS: B
    G3.2

15. Butterflies, like other animals, go through different stages in their life cycle.

    In your **Answer Document**, draw a diagram showing four stages in the life cycle of a butterfly. Label each stage of your diagram. (4 points)

    LS: A
    G3.1

LESSON 10

ECOSYSTEMS

In this lesson, you will learn how plants and animals interact in ecosystems.

— MAJOR IDEAS —

A. An **ecosystem** is made up of all the living and nonliving things in a particular area. The living things in an ecosystem depend on both their physical environment and on one another to survive.

B. A plant or animal can only survive in an ecosystem if its needs are being met. The world has different types of ecosystems, which support various forms of life.

C. Energy and nutrients are cycled through an ecosystem. **Producers** trap energy from the sun. **Consumers** eat producers and other consumers.

D. Living things can affect their environment.

E. Changes to an ecosystem can be *beneficial*, *neutral* or *harmful* to its plants and animals. Some changes may even lead to the **extinction** of some groups.

F. Scientists use **fossils** to provide evidence about the past.

WHAT IS AN ECOSYSTEM?

Have you ever heard of an ecosystem? An **ecosystem** is made up of all the living and nonliving things in a particular area. Every person, animal, plant, and area of land or water belongs to one or more ecosystems. Because they are in the same system, the parts of an ecosystem affect each other in many ways.

A pond's animals and plants form an ecosystem.

A small pond provides a good example of an **ecosystem**. The water, air, sunlight, and mud at the bottom of the pond form its **physical environment**. The moss, pond grass, and small green algae in or around the pond are forms of plant life in this ecosystem.

Insects, snails and other animals living in the pond eat some of these plants to survive. Fish eat snails, insects or smaller fish in the pond. Frogs in the pond eat insects found there. The fish, frogs, and insects all leave behind wastes. Snails, insects, fungi, and bacteria in and around the pond break down these wastes.

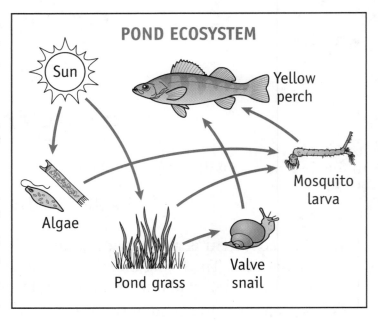

Each of the living things in the pond has its own **habitat**, or special place. Moss grows on the rocks next to the pond. Fish swim in the pond. Frogs jump around the edge of the pond. Each of these are different habitats in the pond's ecosystem.

The different living things in this pond **ecosystem** live in a balanced relationship. Left undisturbed, the pond can continue like this for hundreds or even thousands of years.

APPLYING WHAT YOU HAVE LEARNED

Answer the following questions about a pond ecosystem.

◆ How do the plants in a pond ecosystem depend on their physical environment to meet their basic needs? _____

_____

◆ How do the animals in this ecosystem depend on other living things to meet their basic needs? _____

_____

◆ What are some of the habitats of the animals in this pond ecosystem? _____

_____

Every living thing in an ecosystem must be able to meet its basic needs to survive. These include the need for food, water, shelter, air and waste disposal.

TYPES OF ECOSYSTEMS

The world has many different types of ecosystems. Each type of ecosystem is based on the characteristics of its physical environment, such as climate, how much water there is, and soil type. These characteristics help determine what kind of plants and animals can live in that area. For example, the four types of ecosystems described below are among those found on Earth's land areas.

TEMPERATE FORESTS

Temperate forests develop in regions with 30 to 60 inches of rain each year. The four seasons are marked by moderate temperatures and cool winters. Trees change colors in fall and lose their leaves in winter. There is a wide range of plant and animal life. Insects, spiders, slugs, frogs, turtles, and small and large mammals are common. Animals in this ecosystem must be able to adapt to changing seasons. Some animals move south or hibernate (*sleep*) in the winter.

TROPICAL RAINFORESTS

Tropical rainforests develop in areas where there is ample rainfall and warm temperatures year-round. Large trees cover the forest with their leaves, forming a dense **canopy**. Despite the rapid growth of trees, the soil is actually very thin. Tropical rainforests have a great number of different forms of animal and plant life, with more types of living things than any other type of ecosystem. They are home to more than half of the world's living plant and animal species, including many unique plants, insects, birds, reptiles and mammals.

GRASSLANDS

Grassland areas are found where the climate is drier than a forest but wetter than a desert. There is not enough rainfall to support large numbers of trees. Without trees blocking the sunlight, grasses cover the soil. Large grazing animals like cattle, antelope or bison can survive in these regions by eating the grass.

Cattle are found mainly in grassland areas.

DESERTS

Deserts are regions, such as the Sahara Desert, that receive less than 10 inches of rainfall each year. Deserts have their own special forms of plant and animal life. These plants and animals have adapted to the lack of water and extremes of temperature. A cactus, for example, stores water in its stem. A camel can go for long periods without water. Many insects and reptiles have also adapted well to desert conditions. For example, many only come out at night when it is cooler.

A desert ecosystem in daylight

The same desert ecosystem at night

APPLYING WHAT YOU HAVE LEARNED

Select **one** of the ecosystems you have just read about.

◆ For your selection, list some of the major plants and animals found there.

_____

_____

_____

◆ Describe some of the adaptive characteristics that help these plants and animals survive in their environment. _____

_____

_____

ECOSYSTEM INTERACTIONS

The plants and animals in an ecosystem may interact in different ways.

COMPETITION

Competition is the struggle for survival. Similar types of living things in the same ecosystem will compete for food, water and space. For example, both antelopes and zebras eat grass on the African plains. If there are more zebras and they eat more grass, there will be less grass for the antelopes. If there is less grass for the antelopes to eat, their numbers will decrease.

PREDATORS AND PREY

Many animals in an ecosystem live by eating only plants. However, some animals survive by eating other animals in the ecosystem. For example, lions on the African grasslands cannot live by eating grass. They can only survive by eating other animals. An animal that lives by hunting and eating other animals is called a **predator**. The animal that is hunted is known as its **prey**.

A lynx (predator) chases a rabbit (prey).

PREDATORS

Predators often have special adaptations or characteristics to help them hunt better. These special characteristics include speed, the ability to sneak up while approaching prey, and strong senses of smell, sight and hearing. Predators also follow special patterns of **behavior** to survive in their ecosystems. For example:

★ **Lions** are large members of the cat family. Their large size, powerful claws, and sharp teeth help lions hunt for antelopes, zebras, and other large animals on grasslands of Africa.

★ **Owls** generally hunt at night. In order to do this, they have large eyes to see what they are hunting in the dark. They also have excellent hearing. Their powerful claws and sharp beaks allow them to attack small mammals and birds.

PREY

Prey also have **adaptive characteristics** to help them survive. For example, many types of prey have eyes on the sides of their head. This helps them to see if predators are coming from any direction. Here are other examples:

A lion (predator) captures a zebra (prey).

★ **Horses, antelopes** and **zebras** can run very fast. This helps them to escape when they are in danger of being attacked.

★ **Chameleons** are lizards that use the color of their skin to camouflage themselves. They can even change color. This ability allows them to blend in with their surroundings, such as rocks or leaves, so that predators cannot see them.

★ **Porcupines** have sharp spikes, known as quills, which can hurt a possible predator. When alarmed, a porcupine raises its quills and vibrates them to produce a rattling sound. If that does not scare off the predator, the porcupine charges backwards with its quills up.

Why are sharp quills useful to a porcupine?

COOPERATION

Often there is a cooperative relationship between different plants and animals. In this situation, both plants and animals benefit. For example, bees collect pollen from the flowers of plants. The bees bring the pollen back to their hive to eat. The plant also benefits. As the bees collect pollen, they pass pollen from one plant onto the flowers of other plants. This helps the plants form seeds. Another example is when animals eat fruit and move to another place. They pass out the seeds at the new place in their body wastes. The animals benefit from eating the fruit. The plant benefits because the animals have helped it spread its seeds.

THE FLOW OF ENERGY IN AN ECOSYSTEM

Ecosystems rely on **producers** (*plants*) to bring in energy. This energy then flows through the ecosystem.

PRODUCERS

The plants in an ecosystem produce food from sunlight through **photosynthesis**. They change energy from sunlight into chemical energy, which then enters the ecosystem. All the energy in the ecosystem originally comes from this source.

CONSUMERS

The animals in the ecosystem do not make their own food. They must eat plants or animals to survive. When an animal eats a plant or other animal, it absorbs some of its energy. In fact, all the energy in food can be traced back to plants.

★ **Herbivores**, like cows, eat only plants.

★ **Carnivores**, like lions, eat only other animals.

★ **Omnivores** can eat both plants and animals.

DECOMPOSERS

Some living things in the ecosystem, like ants, worms, fungi, and bacteria, live by breaking down waste products and dead things. These are known as **decomposers**. They put nutrients back into the soil, which are needed by plants.

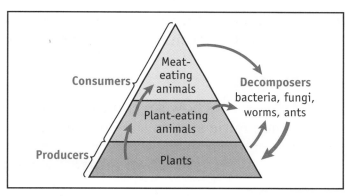

Energy and nutrients are continually recycled in an ecosystem.

APPLYING WHAT YOU HAVE LEARNED

Examine the following list of living things in an ecosystem. Show whether they are producers, consumers, or decomposers:

Organism	Producers, Consumers, or Decomposers?	Organism	Producers, Consumers, or Decomposers?
Worms		Bacteria	
Deer		Mice	
Pine trees		Wheat	
Bears		Algae	
Rabbits		Sparrows	
Ants		Frogs	

FOOD CHAIN

A **food chain** shows the relationship between living things in an ecosystem. It shows *what* eats *what*. Here is a **food chain** from a prairie ecosystem:

In this food chain, rabbits eat the grass. Then coyotes eat the rabbits. In this example, the grass is able to store energy from sunlight through *photosynthesis*. The rabbits take in this energy when they eat the grass. The coyotes absorb some of their energy when they eat the rabbits. In this way, a food chain traces the flow of energy in an ecosystem. The direction of the arrows shows how the energy moves.

Energy in the Food Chain. As you move up the food chain, some energy is lost. One coyote, for example, must eat a large number of rabbits to survive. The rabbits need to eat an even larger amount of grass.

FOOD WEB

A **food web** shows how several living things in an ecosystem interact together. Here is the same prairie system shown as a food web:

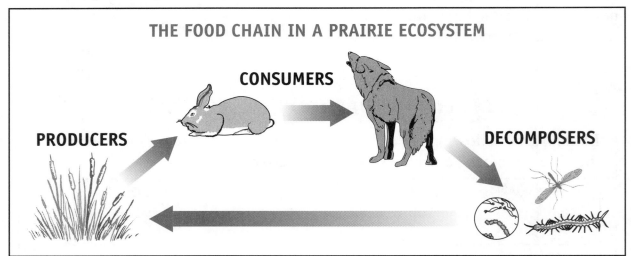

THE FOOD CHAIN IN A PRAIRIE ECOSYSTEM

CONSUMERS

PRODUCERS

DECOMPOSERS

Not only energy, but many **nutrients** are also recycled in an ecosystem. For example, plants have some special chemicals that animals need. When the rabbits eat these plants, they absorb these chemicals. When the coyotes eat the rabbits, they take in the same chemicals. When the prairie grasses, rabbits and coyotes die, their bodies decay. Ants, bacteria and other decomposers break down their remains and return these chemicals to the soil. From the soil, these chemicals are absorbed by the roots of plants. Then the cycle begins all over again.

APPLYING WHAT YOU HAVE LEARNED

Algae produce their own food.

Prawns eat the algae.

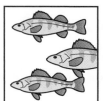

Small fish eat algae and other small fish.

Sharks eat the fish.

Seagulls eat the fish.

Bacteria break down wastes and decaying bodies.

✦ Look at the organisms living together in an ecosystem near the surface of the ocean. Complete the food web below describing this ecosystem:

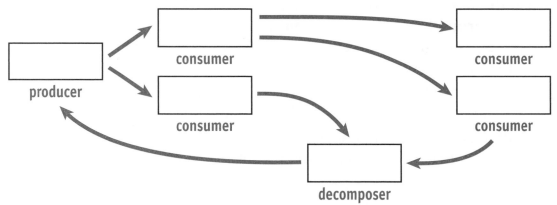

producer

consumer

consumer

consumer

consumer

decomposer

✦ Why is it important that some nutrients in an ecosystem are recycled?

_____

_____

_____

HOW ECOSYSTEMS UNDERGO CHANGE

Ecosystems sometimes change. When a change occurs in an ecosystem, it can affect all members of the ecosystem. If the change is harmful, it may even eliminate some types of plants and animals from the ecosystem.

CHANGES FROM LIVING THINGS

Plants and animals can themselves cause changes to an ecosystem. These changes may be **beneficial**, **neutral** or **harmful** to other living things in the ecosystem. For example:

A beaver using its sharp teeth.

★ **Beavers** are small mammals that live in rivers and streams. They build dams with tree trunks and sticks. The dam creates an artificial pond in the middle of the stream. Other animals may benefit by living in this pond. At the same time, the pond may affect the flow of water in the stream or river.

★ **Earthworms** live beneath the surface of the ground. They dig *burrows*, or tunnels, through the soil and leave their waste material behind. Their activity loosens the soil and makes it more fertile for plants to grow.

★ **Grasshoppers** may eat so many plants that the soil is left unprotected from the wind, sun and rain. This can have a harmful effect on an ecosystem.

★ **Humans** have had both beneficial and harmful effects on the world's ecosystems. People have built dams to prevent flooding and dug irrigation ditches to bring water to new areas. They have brought new plants and animals from one place to another. They have planted trees. However, people have also cut down rainforests, polluted the water and air, and dug mines deep into Earth's surface. These actions have had harmful effects on many of the world's ecosystems.

APPLYING WHAT YOU HAVE LEARNED

Select one example of ecosystem changes from this page and an additional example of your own. For each example, explain how changes to that ecosystem have helped or harmed some of the living things in that ecosystem.

◆ Example from above: _____

_____

◆ Example from your research: _____

_____

CHANGES FROM NATURE

Hurricanes, volcanoes, earthquakes, forest fires, climate changes, meteor impacts and other natural events can also cause changes to an ecosystem.

THE EFFECTS OF CHANGE

Changes to the habitat or ecosystem in which a plant or animal lives can sometimes be helpful. Very often, however, such changes are harmful to the plants and animals in that ecosystem. When changes to an ecosystem take place slowly, plants and animals have time to adapt to those changes. When changes occur rapidly, some plants and animals may not be able to adapt to those changes. These groups lose the ability to meet their basic needs

Huge herds of mammoths once roamed Earth. Now they are extinct.

in the new environment. Those animals or plants die out and become **extinct** — there is no more of that plant or animal left to reproduce. Throughout Earth's long history, many plants and animals have become extinct. Extinct animals include dinosaurs, saber-tooth tigers, and mammoths.

FOSSILS PROVIDE EVIDENCE OF CHANGES

Fossils are impressions in rocks created by the remains of dead plants and animals. Sometimes, a dead plant or animal leaves behind bones, shells, leaves or tracks. For example, a dinosaur might step in mud, leaving its footprint. The mud dries, and sand settles on the mud footprint. The sand and mud harden into different types of rock. A dinosaur's footprints can be seen when the rock is dug up. By examining such fossils, scientists can tell what many

A fossilized fern

plants and animals looked like millions of years ago. For example, scientists can see the leaves of ancient ferns, and the skeletons and imprints left by dinosaurs.

From fossils, scientists have learned about living things that once lived on Earth, but which no longer exist. They can also see that some plants and animals alive today resemble these extinct plants and animals. For example, some reptiles commonly found today resemble dinosaurs.

Fossil of a fish embedded in a rock

From the fossil record, scientists can also often learn about the past environment. For example, the Sahara Desert was once the bottom of a vast sea. Scientists know this because they have found fossils of ancient sea animals in the desert there. Scientists have discovered that the area around the Great Lakes was once buried under icy glaciers, many miles thick.

From fossils, scientists shave also learned that some plants and animals could not survive certain changes that took place on Earth's surface. For example, many scientists believe that a large meteor crashed into Earth's surface about 65 million years ago. The crash created huge clouds

Scientists believe dinosaurs may have become extinct because they failed to adapt to changing conditions.

of dust and smoke that filled the atmosphere and circled Earth. These huge clouds were so thick with dust that they blocked out much of the sunlight. As a result, plants died. The dinosaurs could not get enough to eat. Without food, all of the giant dinosaurs became extinct in a very short period of time.

APPLYING WHAT YOU HAVE LEARNED

◆ How do fossils help scientists learn what happened in the past?

_____

_____

_____

_____

WHAT YOU SHOULD KNOW

You should know that an **ecosystem** is made up of all the living and nonliving things in a particular area.

You should know that the living things in an ecosystem depend on both their physical environment and on one another in order to survive. A plant or animal can only survive in an ecosystem if its needs are being met.

You should know that the world has different types of ecosystems, which support various forms of life. These include forests, grasslands, and deserts.

You should know that energy and nutrients are cycled through an ecosystem. **Producers** (*plants*) trap energy from the sun. **Consumers** (*animals*) eat producers and or other consumers.

You should know that living things can affect their ecosystem. Change to an ecosystem can be harmful to its plants and animals and may even lead to the **extinction** of some groups.

You should know that scientists use **fossils** to provide evidence about the past.

CHAPTER STUDY CARDS

Ecosystems

Ecosystem. All of the living and nonliving things in an area.

★ The characteristics of each plant or animal help it survive and reproduce in its environment.

★ Living things in an ecosystem generally live in a balanced relationship.

★ Animals sometimes **change** their environment to meet their needs. These changes to the ecosystem may be *helpful, neutral* or *harmful* to others.

Flow of Energy in an Ecosystem

★ **Producers.** Plants change sunlight into chemical energy and produce their own food. All food in an ecosystem can be traced back to plants.

★ **Consumers.** These are animals that eat plants or animals for energy.

★ **Decomposers.** Ants, bacteria, and fungi break down wastes and dead plants and animals; they return nutrients to the soil.

★ **Food Chain/ Food Web.** Diagrams that show how energy and nutrients flow through an ecosystem; they show what eats what.

Ecosystem Interactions	Earth's Past

Ecosystem Interactions

Living things in an ecosystem often interact in different ways.

★ **Competition.** Similar things compete for food, water and space.

★ **Predators and Prey.** An animal that survives by hunting and eating other animals is a **predator.** Hunted animals are **prey.**

★ **Cooperation.** Often plants and animals in an ecosystem have a cooperative relationship. They each benefit from their interaction.

Earth's Past

★ Animals and plants that cannot survive become **extinct.**

★ A **fossil** is an impression left by a living thing or its remains, found in rock.

★ Scientists use fossils to tell what living things looked like in the past. Some extinct plants and animals resemble certain living things today.

★ Fossils can also be used to show what an area and its conditions were once like.

CHECKING YOUR UNDERSTANDING

1. Which chain correctly describes the flow of energy in the ecosystem shown to the right?

   A. grass → cow → human
   B. caterpillar → tree → human
   C. cow → grass → human
   D. tree → bird → caterpillar

   LS: B G5.3

HINT To answer this question correctly, you must understand that a food chain shows the flow of energy in an ecosystem. Food is created by plants using the energy of the sun (*photosynthesis*). These plants in turn are eaten by animals (*herbivores*). Other animals may eat these animals (*carnivores*). In this case, the correct order would start with grass. Cows are animals that eat grass. Humans eat beef from cows. Therefore, the correct answer would be **Choice A.**

Now try answering some additional questions on your own:

2. Sheep eat grass and other plants to stay alive, but do not eat animals. What kind of animal are sheep?

   A. carnivores
   B. producers
   C. herbivores
   D. decomposers

   LS: B G5.3

3. An antelope is an animal that eats grass on the plains of Africa. Antelopes are the prey of lions, tigers, and other predators. Which behavior most helps an antelope survive in this ecosystem?

LS: C
G5.5

A. It runs extremely fast.
B. It cannot see tiny objects.
C. It competes with other grazing animals.
D. It cannot smell animals from a great distance.

An antelope.

4. How do decomposers help to keep an ecosystem balanced?

LS: B
G5.3

A. They eat producers.
B. They are food for consumers.
C. They return nutrients to the soil.
D. They make food through photosynthesis.

5. In what order do an owl, acorn, and squirrel form a food chain in the forest?

A.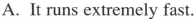

B.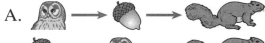

C.

D.

♦ **Examine the Question**
♦ **Recall What You Know**
♦ **Apply What You Know**

LS: B
G5.3

6. If foxes were removed from this food web, which animal population would most likely increase?

A. snakes
B. birds
C. rabbits
D. insects

LS: B
G5.3

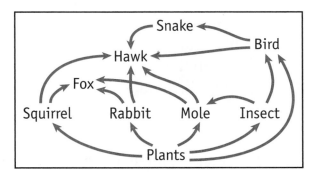

7. Which chain shows the flow of energy in a prairie food chain?

LS: B
G5.3

A.

B.

C.

D.

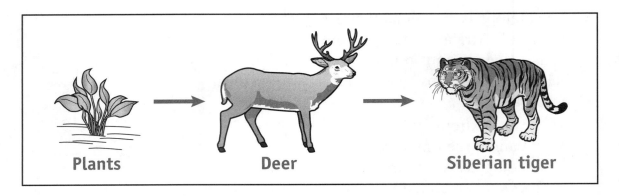

Plants   Deer   Siberian tiger

8. What would be the best title for the diagram?

   A. Animals that are threatened
   B. Energy flow in an ecosystem
   C. Plant and animal differences
   D. All the living things in an ecosystem

   LS: B
   G5.3

9. The African savanna has a warm climate with light rainfall. There are few trees, but enough water to support wild grasses.

   What kind of animals would be best suited to this type of environment?

   A. frogs   C. antelopes
   B. camels   D. polar bears

   LS: C
   G5.4

10. Earthworms dig burrows in the soil, which help loosen the soil. What effect does this activity have on new plants?

    A. neutral
    B. harmful
    C. beneficial
    D. reproductive

    LS: C
    G5.6

Earthworms

11. Plants and animals in an ecosystem interact in a variety of ways.

    In your **Answer Document**, identify two types of interactions that take place in an ecosystem. (2 points)

    LS: A
    G4.5

12. The world has many different types of ecosystems.

    In your **Answer Document**, identify two types of ecosystems.

    Then describe one characteristic of each ecosystem you identify. (4 points)

    LS: C
    G5.4

CONCEPT MAP OF LIFE SCIENCES

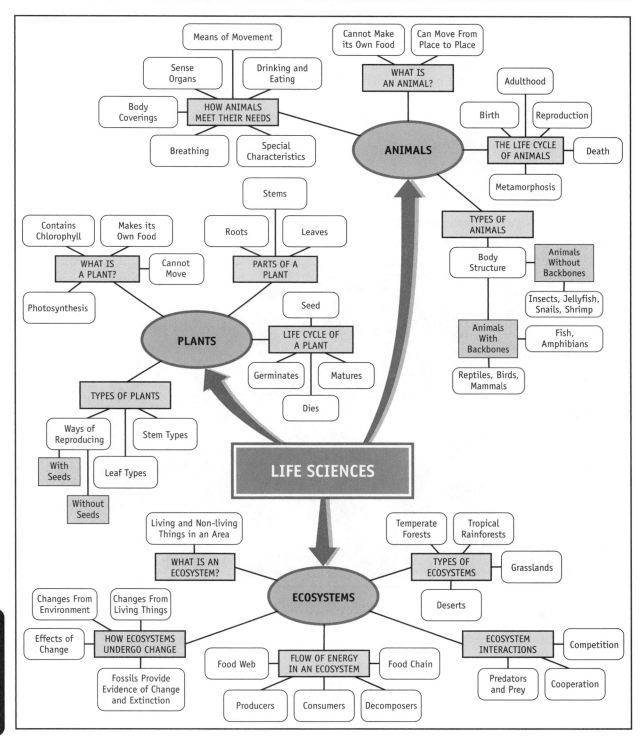

TESTING YOUR UNDERSTANDING

CHARACTERISTICS

Producers	Produce their own energy through photosynthesis
Consumers	Consume other living things for energy

1. Which of the following lists only consumers?

   A. eagles, snakes, mice

   B. grass, rabbits, foxes

   C. acorns, squirrels, owls

   D. rats, chipmunks, trees

   LS: B G5.2

   Use the picture below to answer questions 2 and 3.

Tropical rainforests have high temperatures and heavy rainfall. They are home to many different types of plants and animals.

2. Which part of a bird in the rainforest most helps it to escape predators?

   A. long, thin legs

   B. strong wings

   C. small head

   D. brightly colored feathers

   LS: B G3.2

3. For what resource would consumers in the rainforest most likely compete?

   A. oxygen

   B. sunlight

   C. water

   D. food

   LS: C G5.5

4. Which is common to both temperate forest and rainforest ecosystems?

   A. plant-eating animals
   B. a dense canopy of leaves overhead
   C. trees that lose their leaves in winter
   D. cacti that hold large amounts of water

   LS: B
   G5.1

5. What do plants need to make food?

   A. nitrogen   C. sugar
   B. oxygen   D. sunlight

   LS: B
   G4.2

Use the diagram below to answer question 6.

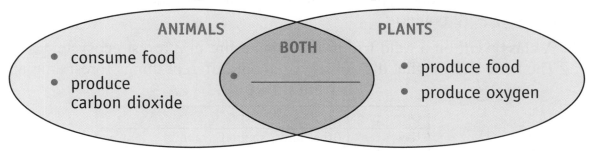

ANIMALS   BOTH   PLANTS
• consume food   • produce food
• produce   _____   • produce oxygen
  carbon dioxide

6. What best completes the blank line in the diagram above?

   A. sleep   C. need water
   B. move about freely   D. have stiff cell walls

   LS: B
   G4.2

7. Which is most likely to contain a fossil?

   A. air
   B. a rock
   D. a plant
   C. rainwater

   ◆ Examine the Question
   ◆ Recall What You Know
   ◆ Apply What You Know

   LS: C
   G3.5

8. Why do the plants in a desert usually have shallow roots that spread out in all directions?

   A. Desert heat makes the roots grow outward.
   B. The roots extend to where it is warmest to grow.
   C. The ground is too hard for the roots to grow any deeper.
   D. Shallow, wide roots help plants capture the scarce rainwater.

   LS: B
   G4.2

9. Which living thing begins life as a seed?

   LS: A
   G4.1

10. The bald eagle eats fish as its main source of food. The fish eats plant life. What conclusion can best be drawn from this information?

A. Most fish are predators.

B. The bald eagle may become extinct.

C. All kinds of food can be traced back to plants.

D. Bald eagles and fish have a cooperative relationship.

LS: B G5.2

11. The world is filled with a large number of ecosystems. Each ecosystem is based on climate, how much water there is, soil type and other characteristics.

In your **Answer Document**, identify one type of ecosystem found in the world.

Then, describe one type of animal or plant life that exists in that ecosystem. (2 points)

LS: C G5.4

12. A class went on a field trip to investigate life in a forest ecosystem. The table shows what they observed about different things forest animals eat:

WHAT EACH ORGANISM EATS

Organism	What It Eats
Hawks	Other birds, squirrels, rabbits
Foxes	Squirrels, rabbits
Squirrels	Nuts, tree buds, insects
Rabbits	Grass, flowers, plant stems

LS: B G5.3

In your **Answer Document**, draw a food web with these plants and animals. When drawing the food web, include the names of four organisms and draw arrows to trace the energy flow between them. (4 points)

CHECKLIST OF SCIENCE BENCHMARKS

Directions. Now that you have completed this unit, place a check (✔) next to those benchmarks you understand. If you are having trouble recalling information about any of these benchmarks, review the lesson indicated in the brackets.

☐ You should be able to distinguish between the life cycles of different plants and animals. [**Lessons 8 and 9**]

☐ You should be able to analyze plant and animal structures and functions needed for survival and describe the flow of energy through a system that all organisms use to survive. [**Lessons 8, and 9 and 10**]

☐ You should be able to compare changes in an organism's ecosystem or habitat that affect its survival. [**Lesson 10**]

UNIT 4

PHYSICAL SCIENCES

In this unit, you will learn about the physical sciences. The **physical sciences** study matter, motion and energy.

You can see matter all around you. Matter is anything that takes up space and has mass.

Energy is what moves or changes matter. Electricity, heat, light and sound are all forms of energy.

Students experiment with different forms of matter.

LESSON 11: MATTER

In this lesson, you will learn about matter and its properties or characteristics. You will also learn how matter can change its state from solid to liquid to gas. Finally, you will learn the differences between physical and chemical changes to matter.

LESSON 12: MOTION AND FORCE

This lesson looks at how matter moves. You will learn that some force is always needed for matter to move or change its motion. You will also learn about different kinds of forces.

LESSON 13: ENERGY

In this lesson, you will learn what energy is — the ability to cause changes in matter. You will also learn about different types of energy, and how one type of energy can change into another.

LESSON 11

MATTER

In this chapter, you will learn about the properties of matter.

— MAJOR IDEAS —

A. **Matter** is anything that takes up space and has mass.

B. Every object made of **matter** can be described by its properties. Some of these properties are **shape**, **color**, **hardness**, **mass**, **magnetism** and the ability to **conduct heat**, **electricity** or **sound**.

C. **Matter** has three states: **solid**, **liquid**, or **gas**. Each state of matter has its own particular physical properties.

D. A **physical change** occurs when matter changes its physical properties without changing its structure.

E. A **chemical change** occurs when two or more substances combine to create new substances with different properties.

WHAT IS MATTER?

Matter is the stuff of the universe. **Matter** is everything that takes up space and has mass. Matter comes in many different shapes and sizes. For example, air and water are both forms of matter. The book you are reading now is matter. Plants and animals are matter. Diamond rings, steam from a boiling tea kettle, sand on a beach, and a piece of chocolate cake are all different forms of matter.

However, not everything is matter. Light and electricity, for example, are not matter. They do not have mass, and they do not take up their own separate space. Your shadow is not matter. Empty space also is not matter.

APPLYING WHAT YOU HAVE LEARNED

Provide two examples of matter and two examples that are not matter.

Matter	Not Matter
1. _____	1. _____
2. _____	2. _____

THE PROPERTIES OF MATTER

Every object made of matter can be described in a variety of ways: its shape, color, texture, mass, volume, hardness, state (*gas*, *liquid*, or *solid*), whether it is magnetic, and how well it conducts heat, sound, and electricity.

Scientists refer to the mass, hardness, state and other characteristics of a piece of matter as its **properties**. A **property** is a **characteristic** or quality that describes the matter. Scientists use these properties to describe and classify different kinds of matter. Let's examine each of these properties more closely.

MASS AND WEIGHT

You might recall from **Lesson 2** that **mass** is *how much* there is of an object — the *amount* of matter in the object. Scientists usually measure mass in **grams** (g) or **kilograms** (kg). They use a **double-pan** or **triple-beam balance** to measure mass.

Although mass and weight are related, they are different. On Earth, an object's mass tells us how much matter there is in the object. The object's **weight** is the amount of force pulling on the object by **gravity**. An object's weight changes if it goes to the moon or some other planet, but not its mass. The chart below shows what you would weigh on Mars or the moon if you weighed 80 pounds on Earth. However, your mass on Mars or the moon *would stay the same*.

A Person's Mass	Earth Weight	Mars Weight	Moon Weight
	80 pounds	30 pounds	13 pounds

MAGNETISM

Another property of some matter is **magnetism**. A **magnet** can attract certain metals, such as **iron**, **nickel**, and **steel**. A magnet will pull pieces of those metals towards it, or even pick them up. Metals attracted to a magnet are **magnetic**. However, many types of metals are not magnetic. For example, a magnet will have no effect on tin, aluminum, copper, gold or silver. A magnet also has no effect on non-metals, like plastic, rubber or wood. Scientists can use magnetism to separate magnetic objects from non-magnetic ones.

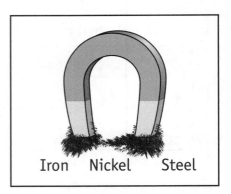

Iron   Nickel   Steel

APPLYING WHAT YOU HAVE LEARNED

✦ A small sliver of iron accidently flew into a child's eye. How could doctors use their knowledge of magnetism to help the child? _____

_____

✦ Which of the following items is magnetic?

☐ a copper penny   ☐ a nickel chain   ☐ a steel toy car
☐ aluminum scissors   ☐ a plastic hanger   ☐ a rubber band
☐ a newspaper   ☐ an iron paper clip   ☐ a gold coin

CONDUCTIVITY

Another characteristic of matter is how well it **conducts** (*carries*) heat, sound or electricity. This property is known as **conductivity**.

HEAT CONDUCTIVITY

Some materials carry heat better than others. In objects made from these materials, heat moves faster from one end of the object to the other. For example, metals conduct heat well. Wood, plastic and rubber do not conduct heat well. For this reason, wood, plastic and rubber are often used for the handle of a pot or pan. They take longer to heat.

SOUND CONDUCTIVITY

Did you know that sound is caused by the vibration of air? Sound vibrations can also pass through liquids and solids. Some materials conduct sound much better than others. For example, metal or wood will conduct sound better than rubber.

APPLYING WHAT YOU HAVE LEARNED

◆ Try this experiment tonight at home. Put your ear against a long piece of wood. Ask a friend to tap the opposite side of the wood. Now cover that end of the wood with a kitchen sponge and have your friend tap it again.

★ Which time was the tapping louder? _____

★ What might explain this? _____

_____

◆ Try this experiment using other items around your house. Which materials conduct sound the best? _____

_____

ELECTRICAL CONDUCTIVITY

Electricity, like heat and sound, can also pass through some materials. Some forms of matter are better conductors of electricity than others. Metals conduct electricity well. Other materials, like rubber or plastic, do not conduct electricity at all. For this reason, wires are usually made of a metal, like copper, surrounded by plastic material that does not conduct electricity. Someone who touches the wire will be protected from the electricity running through the wire, which might otherwise cause a shock.

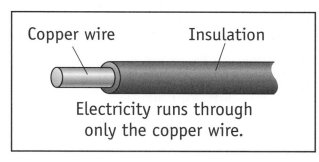

Copper wire   Insulation

Electricity runs through only the copper wire.

APPLYING WHAT YOU HAVE LEARNED

◆ Why is it important for engineers and architects to know which materials are good conductors and which materials are poor conductors? _____

_____

◆ Look around your classroom. Identify three objects that are good conductors of either heat, electricity or sound.

• _____   • _____   • _____

SORTING OBJECTS BY THEIR PROPERTIES

Other properties of matter include its hardness, whether it is shiny or dull, its color, and its state (*whether it is a solid, liquid or gas*).

Scientists are able to use the different properties of materials to describe objects made out of those materials. Scientists can also use those properties to sort objects into groups. For example, scientists can see that paper, glass, metal, and plastic objects look very different. Glass and metal are shiny and smooth. Paper and plastic are less shiny. Paper and plastic feel light to pick up. The same amount of glass or metal is much heavier. Scientists also know that metal conducts heat, sound and electricity well. Plastic, paper and glass cannot conduct electricity.

By looking at the properties of an unknown material, scientists can often tell what it is. They can also tell what materials an object might be made of.

APPLYING WHAT YOU HAVE LEARNED

Jeff examines an unknown object. It has the following properties: it is solid, shiny, and silver in color. When he lifts it, it feels very heavy. He also finds that it can conduct electricity. What would this object most likely be made of?

☐ Wood ☐ Wool ☐ Steel ☐ Sugar

Explain your answer: _____

_____

_____

SEPARATING OBJECTS BY THEIR PROPERTIES

Scientists sometimes use the physical properties of materials to separate them when they are mixed together. For example, a scientist may want to separate a mixture of sand and iron filings (metal shavings). She realizes that the iron filings are magnetic. So she passes a powerful magnet over the pile. The iron filings cling to the magnet, while the grains of sand fall away.

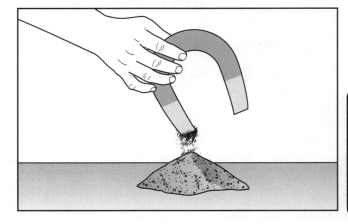

Scientists have also observed that some materials will dissolve in water. Sugar, for example, dissolves in water. If scientists have two powders they need to separate, they may put them in water. One of the materials may dissolve, while the other drops to the bottom of the container.

Scientists may also use a filter or screen to separate materials of different sizes. Small pieces will pass through the filter, while large pieces will not.

APPLYING WHAT YOU HAVE LEARNED

◆ Dolores wants to separate a mixture of sand and sugar. What should she do?

_____

_____

• What physical properties is she using to separate these materials? _____

_____

◆ Pedro wants to separate a mixture of sand from ocean water. What should he do? _____

_____

• What physical properties is he using to separate these materials? _____

_____

THE THREE STATES OF MATTER

Matter can exist in three different **states** or forms: as a **solid**, **liquid** or **gas**. Each of these states has its own physical properties. Matter exists in these different states because it is actually made up of tiny particles. These particles are so small we cannot see them. Scientists believe these particles are constantly moving. It is the motion of these particles that causes the different states of matter.

SOLIDS

In a **solid**, the tiny particles are locked into fixed positions. This gives the matter a **fixed volume** and **shape**. The particles of a solid are still moving, but they vibrate in place. A rock, an ice cube and a diamond are all examples of solids. Solids always hold their own shape. For example, a rock never changes its shape unless something causes it to change.

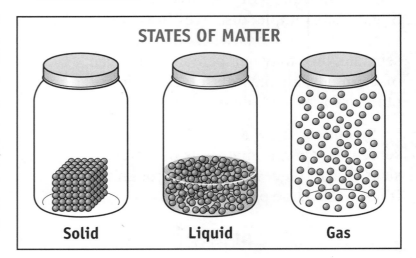

STATES OF MATTER

Solid   Liquid   Gas

LIQUIDS

When heat is applied to a solid, its particles begin to vibrate faster. Eventually, its particles vibrate so fast that they break out of their fixed positions and start to move around each other. The solid **melts** and becomes a liquid. Some examples of liquids at room temperature are water, milk, and mercury. Since the particles of a liquid can move around each other easily, **a liquid does not have a fixed shape**. A liquid takes the shape of the container it is in. For example,

if you pour milk from a bottle into a glass, the milk will take the shape of the glass. Although the shape of a liquid changes, its **volume stays the same**.

GASES

If even more heat is applied to a liquid, its particles will start to move even faster. Eventually, its tiny particles will move so rapidly that they will spread out in all directions as a **gas**. When this happens, the liquid **boils** and turns into a gas. A gas has **no fixed shape** and **no fixed volume**. It will fill up whatever space it has. Some examples of common gases at room temperature are oxygen and carbon dioxide.

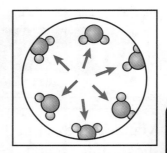

Water and some other liquids have a special characteristic. Even if it does not boil, an open container of water will **evaporate** into the air. When water evaporates, it turns from a liquid into a gas. In addition, a few solids, like dry ice, can go directly from a solid to a gas.

APPLYING WHAT YOU HAVE LEARNED

✦ How do solids, liquids and gases differ from each other? Complete the following chart by filling in the characteristics of each.

Characteristic	Solid	Liquid	Gas
Movement of Particles	Particles vibrate in place		
Shape			No fixed shape
Volume		Has a fixed volume	

PHYSICAL AND CHEMICAL CHANGES

In the last two sections of this lesson, you learned about the physical properties of matter. In this section, you will learn how matter sometimes changes its properties. For example, a pot of water might boil. This is a **physical change**.

Another type of change occurs when a log burns in a fireplace. The log catches fire and burns with a crackling flame. The fire produces beautiful colored flames and relaxing heat. After it has finished burning, the log is gone. In its place is a heap of ashes. This is an example of a **chemical change**.

APPLYING WHAT YOU HAVE LEARNED

✦ From these examples, can you explain how physical and chemical changes are different? _____

_____

PHYSICAL CHANGES

In a **physical change**, matter changes one or more of its physical properties. For example, the water in a pot is liquid. After the water boils, it becomes a gas. But this matter is still water. It has not changed its basic structure. In fact, **changes of state** (*boiling, condensing, freezing* and *melting*) are always physical changes. Another example is iron. Iron is a very strong and hard metal. You cannot scratch it easily. However, if you heat an iron rod to high temperatures, it becomes softer. A blacksmith can shape it into a sword, a horseshoe, or a tool.

Physical changes are **reversible**. If the blacksmith cools down the iron, it will become strong and hard again. Water that has been boiled can also be cooled down. Drops of moisture will appear as the water vapor turns from a gas back into a liquid.

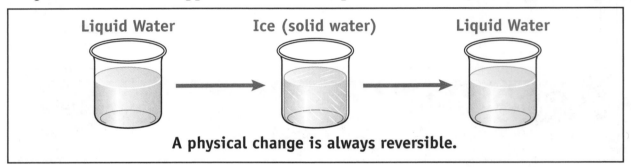

Liquid Water   Ice (solid water)   Liquid Water

A physical change is always reversible.

CHEMICAL CHANGES

A chemical change is different from a physical change. In a **chemical change**, matter changes its structure. One type of matter combines with one or more other types of matter to form something completely different. For example, oxygen gas and hydrogen gas might be combined in a laboratory. When these two gases are brought together under the right conditions, a chemical change occurs. Particles of oxygen and hydrogen join together to form a completely new substance — water. Water has properties that are different from either oxygen or hydrogen gas. At room temperature, water is a liquid, while oxygen and hydrogen are gases.

A chemical change results in the creation of new substances.

The materials formed by a chemical change also have new chemical properties, different from those of the original materials. A **chemical property** is the ability of one material to combine with others in a chemical reaction. Hydrogen, for example, can burn. Water cannot.

APPLYING WHAT YOU HAVE LEARNED

A scientist mixed together two common household materials — vinegar and baking soda. The scientist poured one cup of vinegar into a container with half a cup of baking soda. The mixture began to bubble. Soon, bubbles were pouring out of the container. As the vinegar and baking soda combined, they produced carbon dioxide gas.

◆ Was this a physical or chemical change? _____

Explain your answer: _____

_____

In Unit 2, you learned about photosynthesis. In photosynthesis, plants combine water, carbon dioxide gas, and sunlight to produce oxygen and a form of sugar.

◆ Is this a physical or chemical change? _____

Explain your answer: _____

_____

WHAT YOU SHOULD KNOW

☐ You should know that **matter** is anything that takes up space and has mass.

☐ You should know that every object made up of **matter** can be described by its properties. Some of these properties are mass, hardness, magnetism, and the ability to conduct heat, electricity or sound.

☐ You should know that matter has three states: **solid**, **liquid**, or **gas**. Each state of matter has its own particular physical properties.

☐ You should know that a **physical change** occurs when matter changes its physical properties without changing its structure.

☐ You should know that a **chemical change** occurs when two or more substances combine to create new substances with different properties.

CHAPTER STUDY CARDS

Properties of Matter

Every type of matter has certain **properties**.

★ **Mass.** How much there is of an object — the amount of matter usually measured in grams (g) and kilograms (kg).

★ **States of Matter.** The three states of matter are **solid**, **liquid**, and **gas**.

★ **Magnetism.** A force of attraction between a magnet and some metals.

★ **Conductivity.** How well something carries heat, sound, or electricity.

★ **Other Properties.** Other properties of matter include how hard or soft it is, how it looks, sounds, tastes, or smells.

Sorting Objects by Their Properties

★ Scientists use different properties of materials to describe objects made out of those materials.

★ By looking at the properties of an unknown material, scientists can sometimes tell what material that object is made of.

★ Scientists also use the physical properties of materials to separate them when they are mixed together.

States of Matter

Matter can exist in one of three forms, based on the motion of its particles:

★ **Solid.** It has a **fixed volume** and **shape**.

★ **Liquid.** A liquid has **no fixed shape**: it takes the shape of whatever container it is in; however, it has a **fixed volume**.

★ **Gas.** Its particles move in all directions. A gas has **no shape** and **no fixed volume**.

Physical and Chemical Changes

★ A **physical change** occurs when matter changes one or more of its physical properties.

• Changes of state are always physical changes.

• Physical changes are reversible. An example would be ice melting.

★ A **chemical change** occurs when one type of matter combines with another to become something that has a completely different structure. An example: burning of wood.

CHECKING YOUR UNDERSTANDING

1. Which of the illustrations below shows a liquid changing into a gas? (PS: A G4.1)

 A.  B.  C.  D.

HINT

To answer this question, you need to understand the different states of matter — solid, liquid, and gas. Picture A shows a solid cooking; Picture B shows a liquid boiling; Picture C shows popcorn popping over a flame; Picture D shows a liquid being poured onto a frying pan. Therefore, **Picture B** is the correct choice. When water boils, it changes its state from a liquid into a gas.

Now try answering some additional questions on your own:

2. Which of the following objects would be attracted to a magnet?

    A. paper bag   C. iron nail

    B. rubber ball   D. copper penny

> PS: B
> G4.3

3. Matter can change its form. What happens when water freezes?

    A. A liquid becomes a gas.

    B. A gas becomes a liquid.

    C. A liquid becomes a solid.

    D. A solid becomes a liquid.

> ◆ **Examine the Question**
> ◆ **Recall What You Know**
> ◆ **Apply What You Know**

> PS: A
> G4.1

4. In this picture, the nail is pulled by object X, but the wood block next to it does not move. What property does object X have that attracts the nail?

    A. density

    B. solid state

    C. magnetism

    D. high temperature

> PS: B
> G4.3

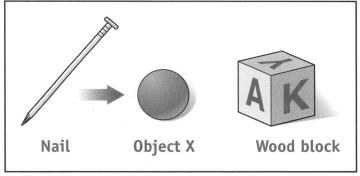

Nail   **Object X**   **Wood block**

The picture above shows a nail being attracted by object X

5. A student takes an ice cube tray out of the freezer. What happens when the ice cubes melt?

    A. liquid $\rightarrow$ gas   C. solid $\rightarrow$ liquid

    B. liquid $\rightarrow$ solid   D. gas $\rightarrow$ solid

> PS: A
> G4.1

6. Which is an example of a chemical change?

    A. A scientist melts ice to create water.

    B. A scientist shapes melted glass into a cylinder.

    C. A scientist grinds an iron rod to make iron filings.

    D. A scientist mixes chlorine gas and sodium to make salt.

> PS: A
> G4.2

7. Iron filings are dark metal shavings. When iron filings are mixed with sand, they create a grayish powder. Once these materials have been mixed together, what would be the best way to separate them?

   A. Heat the mixture.
   B. Pour the mixture through a filter.
   C. Put the mixture in water and stir it.
   D. Use a magnet to pull out the iron filings.

   PS: B
   G4.3

8. Which is an example of matter changing from a liquid to a gas?

   A. A tire loses air.
   B. A glass dish breaks.
   C. A pot of soup boils.
   D. A piece of plastic burns.

   ◆ **Examine the Question**
   ◆ **Recall What You Know**
   ◆ **Apply What You Know**

   PS: A
   G4.1

9. Which is an example of a physical change?

   A. A piece of paper burns.
   B. A steel rod is heated until it melts.
   C. Hydrogen and oxygen gases combine to make water.
   D. A plant turns water and carbon dioxide into sugar and oxygen.

   PS: A
   G4.1

10. Which of these is the best conductor of heat?

    A. a rubber glove   C. a copper pot
    B. a plastic spoon   D a wooden handle

    PS: B
    G4.3

11. What process is illustrated in the diagram on the right?

    A. boiling
    B. melting   *PS: B*
    C. freezing   *G4.4*
    D. burning

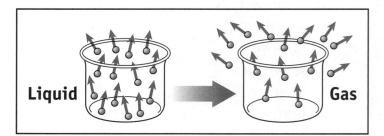

12. In an experiment, sugar and iron filings are mixed together. What would be the fastest way to separate the iron filings from the sugar?

    PS: B
    G4.3

    A.
    • stir the mixture in water
    • pour the wet mixture through a filter

    C.
    • put the mixture in a beaker
    • heat the beaker over a hot plate

    B.
    • pour the mixture on a piece of paper
    • use a pin to separate the filings

    D.
    • stir the mixture in water
    • boil the mixture

LESSON 12

MOTION AND FORCE

In this lesson, you will learn about motion and force and how they are related.

— MAJOR IDEAS —

A. An object's **location** is based on where it is in relation to other things.

B. **Motion** consists of speed and direction.

C. When **force** is applied to an object, it changes its motion.

D. **Gravity**, **magnetism**, **collision** and **friction** are all examples of forces that can affect an object's motion.

DESCRIBING MOTION

Motion occurs when an object changes its location. If you watched a race, you would see runners move from one location to another.

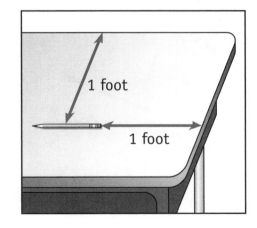

LOCATION

Location is the position of an object in relation to other things. For example, you can indicate the location of your pencil based on its distance from the edges of the desk. Your pencil might be one foot from the right side of your desk and one foot from the back edge of your desk.

DISTANCE

Distance is the length that something moves. For example, you might move your pencil to the very back edge of your desk. In this case, your pencil has moved a distance of one foot.

SPEED

Speed is how *fast* an object moves. It is the **distance** the object travels **in a unit of time**. For example, a car may travel 10 kilometers an hour (10 km/h). This means that every hour, the car travels 10 kilometers. How many kilometers would that car travel in two hours? In three hours? Fill in the chart to the right.

Time	Distance Traveled
1 hour	★ 10 kilometers
2 hours	★ _____
3 hours	★ _____
4 hours	★ _____
5 hours	★ _____

You can turn this same information into a line graph. Complete the graph below:

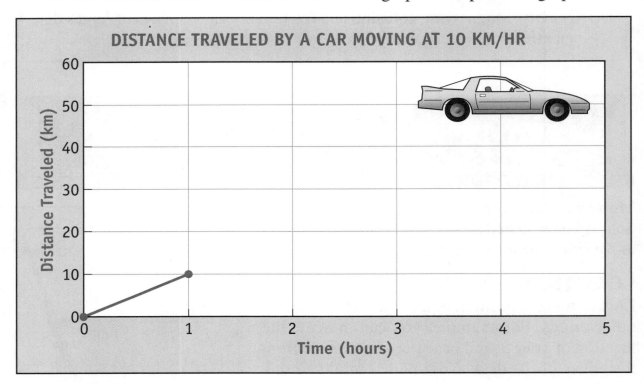

DIRECTION

Another thing to consider when looking at motion is **direction**. If a car travels in the same direction for two hours at 50 km an hour, it will end up 100 km from its original location. But if it travels for one hour and then turns back, at the end of two hours it will be back at the starting point.

APPLYING WHAT YOU HAVE LEARNED

✦ A car travels for three hours at 80 km/hour. After three hours, it turns right. It drives 60 km/hour for three hours. How many km has it traveled? _____ What was the average speed of the car? _____

✦ Create a diagram tracing the movement of this car.

TYPES OF FORCE

What causes objects to move? Anything that causes an object to move is known as a **force**. There are four important types of forces you should know:

GRAVITY

Gravity is the force that pulls objects to the ground on Earth. You learned in the last lesson that the force of gravity is responsible for an object's **weight**. Scientists have learned that gravity not only exists on Earth. It is a force of attraction between any two objects. The force of gravity can act across space. Two objects do not need to be touching to be pulled together by gravity. For example, the moon is attracted to Earth by gravity. Gravity also attracts Earth to the sun. The larger the mass of each object, the stronger the force of gravity will be. That is why objects with greater mass weigh more on Earth. They are pulled to Earth with greater force.

APPLYING WHAT YOU HAVE LEARNED

✦ Which object is pulled to Earth by a greater gravitational force: a paper clip or a wooden table? _____ Explain your answer. _____

_____

✦ Explain why the weight of an object is less on the moon than on Earth.

_____

MAGNETISM

In the last lesson, you learned about magnetism. **Magnetism** is a force of attraction between a magnet and many metals. Like gravity, it is a **non-contact force**: a magnet can pull a metal object towards itself even though it is not touching it. With a magnet, you can pick up paper clips or small nails made of steel. The paper clips or nails are attracted to the magnet. They move towards it and will stick to it. Every magnet has two sides, or poles. The North Pole of one magnet will attract the South Pole of another magnet. However, two North Poles or two South Poles will repel, or push away, one another.

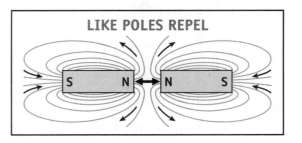

LIKE POLES REPEL

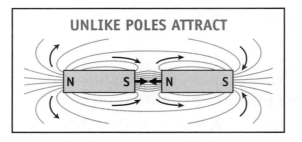

UNLIKE POLES ATTRACT

COLLISION

Collision occurs when one object crashes into, or *collides with* another object. It is a **contact force**. The motion of the moving object transfers to the object it strikes. This will cause an object at rest to start moving, or it will affect the direction or speed of a moving object that is hit.

Suppose that two smooth, hard balls of the same weight collide:

A BALL COLLIDES INTO A BALL AT REST:

1. A moving ball hits a ball at rest.

2. The ball at rest starts to move in the same direction as the moving ball.

A BALL COLLIDES INTO A BALL MOVING IN THE SAME DIRECTION:

1. A faster moving ball hits a slower ball moving in the same direction.

2. The second ball speeds up.

A BALL COLLIDES INTO A BALL MOVING IN THE OPPOSITE DIRECTION:

1. A moving ball hits another ball moving towards it.

2. The balls will each bounce back in the direction they came from.

A BALL COLLIDES INTO A BALL MOVING IN A DIFFERENT DIRECTION:

1. A moving ball hits another ball moving in a different direction.

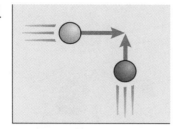

2. The second ball will change its speed and direction.

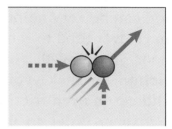

FRICTION

Friction is the force that occurs when two things rub against each other. The rubbing of their surfaces slows their motion. For example, if you rub your two hands together, you create friction. On Earth, moving objects experience friction from the air and the ground. The force of this friction slows down moving objects and eventually leads them to stop unless some other force is applied to keep them going.

APPLYING WHAT YOU HAVE LEARNED

When a car moves ahead, the force from the wheels moving the car forward is greater than the opposing force of friction — the rubbing of the tires against the roadway. Give two other examples from everyday life where

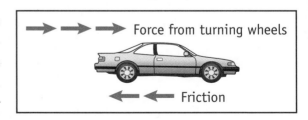

Force from turning wheels

Friction

some force is applied and leads to changes in an object's speed or direction.

A. _____

B. _____

WHAT FRICTION DEPENDS ON

The amount of friction depends on the type of surfaces that are rubbing against each other. When surfaces are smooth, they move against each other more easily. Friction is greater between rough surfaces. In order to reduce friction in the moving parts of machines, scientists and engineers use oil and other *lubricants*. These lubricants make the surfaces smoother.

IS FRICTION GOOD OR BAD?

Friction is not all bad. In fact, friction can be very helpful. If you have ever tried to run on a wet floor, you know that too little friction can make you slip and slide. You need friction to get a grip in order to move. Without friction, we would be unable to walk or climb stairs. Machines would not work. On the other hand, too much friction will quickly wear down a machine's moving parts.

APPLYING WHAT YOU HAVE LEARNED

✦ Which has more friction, a roadway before or during a heavy rainfall? Explain your answer. _____

_____

_____

_____

PREDICTING CHANGES IN MOTION

How do each of these types of forces cause changes to the motion of objects: gravity, friction, magnetism, and collision? In all cases, how much the motion of an object will change depends on the amount of force applied to it and the size of the object. An object with greater mass requires more force to change its motion than an object with less mass.

IN GENERAL, THESE FOUR TYPES OF FORCES CHANGE MOTION IN THESE WAYS:

Type of Force	How It Changes Motion
Gravity	Pulls objects towards Earth. Gravity increases with the mass of the object.
Magnetism	Pulls many types of metals towards a magnet.
Collision	Transfers motion from one object to the object it strikes. A collision can change the speed or direction of motion, or make an object at rest move. (See the diagrams on pp. 140–141.)
Friction	Slows down motion.

APPLYING WHAT YOU HAVE LEARNED

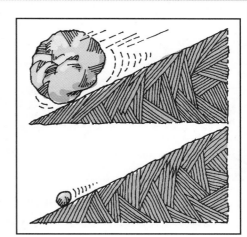

◆ A large boulder is rolling down a hill. How much force do you think would be needed to stop it? Explain your answer.

_____

_____

_____

◆ A small pebble is rolling down the same hill at the same speed as the boulder. Is more or less force needed to stop the pebble than the boulder? Explain your answer.

_____

_____

WHAT YOU SHOULD KNOW

☐ You should know that an object's **location** or **position** is based on where it is in relation to other things.

☐ You should know that **motion** consists of speed and direction.

☐ You should know that when **force** is applied to an object, it changes its motion.

☐ You should know that **gravity**, **magnetism**, **collision** and **friction** are examples of forces that can affect an object's motion. Gravity and magnetism are **non-contact forces**. Collision and friction are **contact forces**.

CHAPTER STUDY CARDS

Motion

★ **Motion** is the act of changing the position of an object. It has speed and direction.

★ **Speed.** Speed measures the distance an object travels in a given amount of time, often measured in km/h.

★ **Direction** is the path or route an object takes.

★ **Force** is what pushes or pulls an object to make it change its speed or direction. An object with greater mass requires more force to change its motion than one with less mass.

Types of Force

A. Non-Contact Forces:

★ **Gravity** is the force pulling objects to Earth. The force of gravity increases with the mass of the object.

★ **Magnetism** is the force of attraction between a magnet and many metals.

B. Contact Forces:

★ **Collision** is the force with which a moving object strikes another.

★ **Friction** is the force created by the rubbing of two surfaces.

CHECKING YOUR UNDERSTANDING

1. If the person in the picture lets go of the rope, the weight (w) will fall to the ground.

   What force is acting to pull the weight to the ground?

   A. friction
   B. gravity
   C. collision
   D. magnetism

 To answer this question correctly, you have to be able to identify different types of forces. When the person lets go of the rope, gravity is the force that pulls the weight to Earth. The best answer is **Choice B.**

Now try answering some additional questions on your own:

2. A student measures the time it took her to run 50 meters. What is she able to calculate using both her distance and time measurements?

   A. her force   C. her speed
   B. her weight   D. her direction

3. Which tools are needed to measure the speed of a rolling ball?

   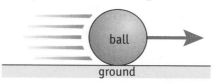

   A. stopwatch, ruler
   B. spring scale, ruler
   C. thermometer, balance
   D. thermometer, stopwatch

   PS: C
   G3.2

4. Jasmine watches cars race around a track. What information does she need to describe the motion of the cars?

   A. the shape of the cars
   B. the speed of the cars
   C. the weight of the cars
   D. the temperature of the tires

   ♦ **Examine the Question**
   ♦ **Recall What You Know**
   ♦ **Apply What You Know**

   PS: C
   G3.2

5. What force slows down a baseball rolling on a field?

   A. heat   C. friction
   B. light   D. magnetism

   PS: C
   G3.4

6. Libby went on a trip with her family. In her log, she recorded the time of day and how far her family had traveled since they started.

Time of Day	Total Distance Traveled
8:00 AM	0 km
9:00 AM	30 km
10:00 AM	60 km
11:00 AM	90 km

   What speed is Libby's family traveling in kilometers/hour?

   A. 0 km/h   C. 30 km/h
   B. 10 km/h   D. 90 km/h

   PS: C
   G3.2

7. The graph to the right shows the movement of a car during a four-hour trip. What was the speed of the car, if it traveled at the same speed for all four hours?

   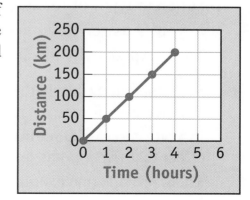

   A. 0 km/h
   B. 50 km/h
   C. 100 km/h
   D. 200 km/h

   PS: C
   G3.2

8. Which of the following is most likely to increase friction?

   A. wax on wood floors
   B. snow on a roadway
   C. wheels on roller skates
   D. grooved rubber soles on shoes

   PS: C
   G3.4

9. In order to describe an object's motion, what is needed besides its speed and its position?

   A. its color
   B. its height
   C. its weight
   D. its direction

   ◆ **Examine the Question**
   ◆ **Recall What You Know**
   ◆ **Apply What You Know**

   PS: C
   G3.2

10. Mr. James mops the floors in the school each afternoon. When he finishes mopping, he always puts up a sign stating, "Warning: Floor Slippery When Wet." Which force is reduced when the floor is wet?

    A. heat
    B. friction
    C. collision
    D. magnetism

    PS: C
    G3.4

11. Keisha lives next door to Mr. Murphy and three houses away from Ms. Riley. What does this statement identify?

    A. her speed
    B. her force
    C. her location
    D. her direction

    PS: C
    G3.1

    PS: C
    G3.3

12. Different types of forces can affect how an object moves.

    In your **Answer Document**, identify one non-contact force.

    Then, explain why this is considered a non-contact force. (2 points)

13. Some people think that friction is good, while others believe that friction acts as a negative force.

    In your **Answer Document**, identify one positive and one negative effect that friction has in daily life.

    Then, provide one example to illustrate each type of effect. (4 points)

    PS: C
    G3.4

LESSON 13

ENERGY

In this lesson, you will learn about different types of energy.

— MAJOR IDEAS —

A. **Energy** has the ability to move or change matter. There are different types of energy. These include **thermal energy** (*heat*), **electricity**, **light** and **sound**.

B. The **thermal energy** in an object changes as the motion of its particles speeds up or slows down. **Temperature** measures the thermal energy of an object.

C. **Electrical energy** can flow through a **circuit**. Electrical energy can produce thermal energy, light, sound and magnetism.

D. **Light** travels through space by waves in a straight line. When light energy hits an object, it is either absorbed, transmitted, refracted (*bent*) or reflected.

E. **Sound** is caused by **vibrations**. Changing the speed of the vibrations will change the pitch of the sound. Like light, sound can be absorbed, transmitted, or reflected.

ENERGY

In **Lesson 12**, you learned about force. Force is always created by some kind of energy. You cannot always see energy, but you can often feel it. Some days you may feel more "energetic" than others. You feel like doing things. **Energy** is the ability to cause changes in matter. There are many kinds of energy. These include:

| Thermal Energy | Electricity | Light | Sound |

THERMAL ENERGY

Thermal energy, also known as *heat energy*, is caused by the movement of the tiny particles that make up matter. As the particles in an object move around more quickly, the object heats up. Its thermal energy increases. **Temperature** measures the thermal energy of an object — how fast its tiny particles are moving. For example, water particles are moving faster at 100°C than at 50°C.

The thermal energy of an object may change. Different actions can change the speed of its particles. **Rubbing**, for example, can make an object warmer. The rubbing causes its particles to move faster. **Bending** a piece of metal back and forth also causes its particles to speed up. This causes an increase in temperature.

Conduction. Thermal energy can pass from an object to neighboring objects in direct contact. This process is known as **conduction**. The moving particles in the object bump into the particles of neighboring objects and speed them up.

In conduction, thermal energy always moves from warmer objects to cooler ones. For example, if you stirred a pot of hot soup on the stove with a metal spoon, after a few minutes the metal spoon would heat up. The thermal energy passed from the hot soup to the spoon. Because thermal energy transfers from one object to another, an object's thermal energy can be increased by **heating**. Thermal energy from a hot flame transfers to a pot over the flame.

Thermal Clothing. Thermal socks and other clothing items are made of materials that are poor conductors of heat. This slows down the transfer of thermal energy. Thermal socks help trap the heat our bodies create to keep us warm on cold days.

APPLYING WHAT YOU HAVE LEARNED

Three boxes with different temperatures are placed together:

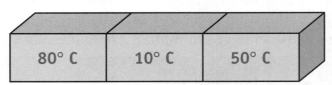

◆ How will thermal energy transfer among these three boxes? _____

_____

ELECTRICAL ENERGY

Electricity is another form of energy. It is made by fast-moving, charged particles. Each of these particles carries some energy. Electricity has many special properties. Electricity can flow easily through many types of materials. Most metals are good conductors of electricity. However, not all materials conduct electricity. Wood and rubber do not conduct electricity.

ELECTRIC CIRCUITS

Electricity can flow in a circuit. A typical circuit has several parts:

A source that provides electricity	Something that uses electricity	Wires to carry the electricity

A **battery** is one source that produces electrical energy. Wires connected to a battery can carry electricity. However, the electricity will only flow if the circuit is **complete**. This provides a continuous path for the energy to move around. Scientists call this a **closed circuit**.

For example, imagine a simple circuit with a light bulb at one end and a battery at the other. (***Figure 1***). Electricity leaves one side of the battery and moves through the wire to the light bulb. The electricity then moves through the bulb and causes it to light up. The electrical current then continues moving along the wire back to the battery.

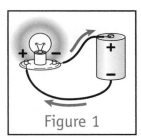
Figure 1

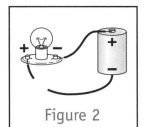
Figure 2

If the circuit is cut at any point, the electricity flowing through the circuit will stop. The circuit is incomplete. Since the light bulb has no electricity flowing into it, it will go out (***Figure 2***).

The wires must also be connected to each side of the battery. If both ends of the wire are connected to the same side of the battery, the electricity will not leave the battery. The bulb will not receive any power and will not light up (***Figure 3***).

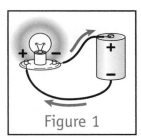
Figure 3

If both ends of the wire are connected to the same side of the light bulb, the electricity will go around the circuit without going into the light bulb. Again, the bulb will not light up (***Figure 4***).

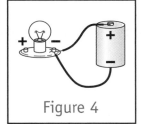
Figure 4

APPLYING WHAT YOU HAVE LEARNED

Each circuit below has a light bulb, a battery, and wires.

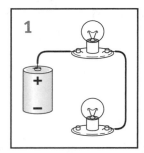
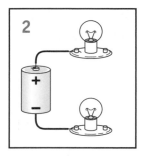
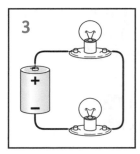
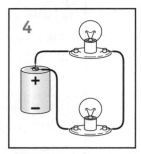

✦ Which circuit will light its bulbs? _____

✦ Explain why the other bulbs will not light up. _____

_____

_____

OTHER PROPERTIES OF ELECTRICAL ENERGY

Electrical energy also has several other important properties. These properties include *magnetism*, *heat*, *light* and *sound*.

★ **Magnetic Force.** When electricity runs through a wire, the wire becomes **magnetic**. A compass, for example, will point to a wire with electricity. A temporary magnet made by coiled wires wrapped around metal is called an **electromagnet**. When the electricity is turned on, the metal becomes magnetic. When the electricity is off, it is not magnetic This property of electricity makes it possible to produce electric motors. They use electromagnetic force to turn.

★ **Thermal Energy.** When electricity runs through some materials, it makes them hot. Irons, waffle makers, toasters, and electric heaters all produce thermal energy when electricity runs through them.

★ **Light.** When electricity runs through some materials, it makes them so hot they glow. Because of this property, electricity makes light bulbs work. Electricity also makes flashes of lightning in the sky.

★ **Sound.** Electricity is also capable of making sound. Televisions, radios, and stereo speakers use electricity to make sound. In nature, electricity causes the loud sounds of thunder.

APPLYING WHAT YOU HAVE LEARNED

Examine the electrical appliances illustrated below.

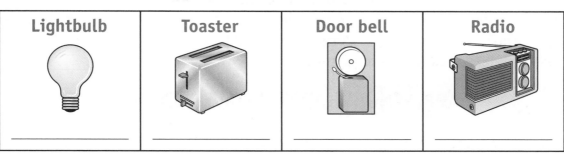

Lightbulb	Toaster	Door bell	Radio
___	___	___	___

✦ Describe what kind of energy each appliance produces. Then explain how electricity flows through each item. _____

_____

LIGHT

Light is a third form of energy. Light comes from the energy given off when the tiny particles that make up matter collide or move extremely fast. Light energy can move through water, air, glass and many other materials. Light energy can even move through empty space. **Solar energy** is a form of light energy. This light energy from the sun passes through millions of miles of space to reach Earth.

Light usually travels by waves in straight lines outward from its source. Although light can pass through many materials, it cannot pass through others. Some materials, such as black construction paper, absorb light. Other materials **reflect** or **refract** light.

REFLECTION

Hard, shiny surfaces reflect light. When light is **reflected**, it simply bounces off the surface it strikes. For example, if the beam of a flashlight hits a mirror, the beam will bounce back off the mirror. As the illustration shows, the light will bounce off the mirror with the same angle it came, but in the opposite direction.

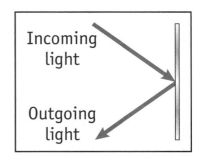

Incoming light

Outgoing light

REFRACTION

Some materials actually bend light. This is known as **refraction**. Refraction often happens when light passes through one substance into another.

Not all the light rays travel at the same speed in the second substance. This makes the light appear to bend. You can see this when you put a pencil in a glass of water. The pencil seems to bend. This property of light is very useful. It makes it possible to make **lenses** — curved pieces of glass that bend light. Lenses are used to make eyeglasses, telescopes, and microscopes.

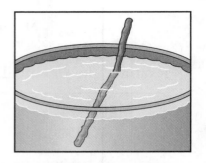

APPLYING WHAT YOU HAVE LEARNED

✦ How will these objects affect light? More than one box may apply.

Object	Transmits Light	Absorbs Light	Reflects Light	Refracts Light
Water in a calm lake				
Pair of eyeglasses				
Shiny silver car				
Pair of blue jeans				

✦ Explain what happens in each of the following three examples:

⭐ **Figure A.** Light shines directly into a glass of water. The light strikes the water straight on. Explain what will happen to the light.

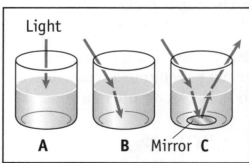

_____

_____

⭐ **Figure B.** Light shines into the glass of water at an angle. As the light enters, explain how the beam of light appears at the surface of the water.

_____

⭐ **Figure C.** A mirror is placed at the bottom of a glass of water. Light shines into the glass at an angle. Explain what happens to the light when it hits the mirror at the bottom of the glass.

_____

SOUND

Sound is created by **vibrations**. For example, when a person pulls a guitar string, the string begins to vibrate. The energy from the vibrating string causes the air to vibrate. These vibrations spread out in waves. Our ears are very sensitive to these vibrations. The energy from these vibrations passes to our ears, and we hear these vibrations as sound. If the speed of the vibration changes, the **pitch** of the sound will also change. More vibrations per second create a higher pitch. Fewer vibrations make the pitch lower.

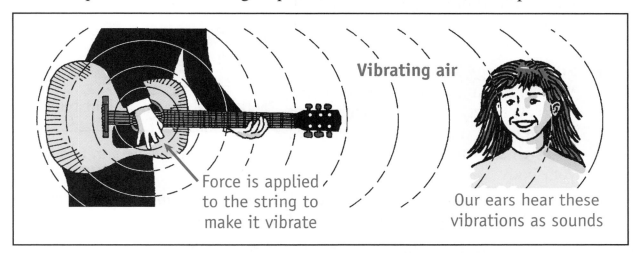

Vibrating air

Force is applied to the string to make it vibrate

Our ears hear these vibrations as sounds

Characteristics of Sound. Most sounds we hear travel through the air. But sound can also travel through many other forms of matter, including liquids and solids. Sound actually travels faster through many solids than it does through the air. Like light, sound can be transmitted, reflected, or absorbed. If sound waves hit a hard, smooth surface, they may just bounce back. This reflected sound is known as an **echo**. Curtains, carpets, and thick fabrics are able to **absorb** sound. Acoustic engineers use these characteristics of sound to design auditoriums and concert halls.

APPLYING WHAT YOU HAVE LEARNED

1. If you put your ear on the wall, it is sometimes possible to understand what is being said in the next room. Explain why. _____
   _____

2. Can sound travel through empty space? _____ Why or why not? _____
   _____

3. What materials would you use to design a quiet bedroom in your house?
   _____

ENERGY CAN CHANGE ITS FORM

Energy can actually change its form. Now that you know about the different kinds of energy, you can see how this works:

★ **Electrical Energy → Light Energy.** Electricity runs through a material that gets very hot. It can get so hot that it makes the thin wire in a light bulb glow. In this example, electrical energy turns into light energy.

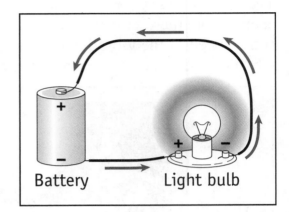
Battery   Light bulb

★ **Solar Energy → Thermal Energy.** We use the sun's energy in different ways every day. If you go to the beach and lie in the sun, after a few minutes you will start to feel hot. The energy that comes from the sun is heating your body.

★ **Solar Energy → Chemical Energy.** Through photosynthesis, plants change energy from sunlight into the chemical energy stored in food.

WHAT YOU SHOULD KNOW

☐ You should know that **energy** is an ability to change or move matter. There are different types of energy. These include thermal energy, electricity, light and sound.

☐ You should know that the **thermal energy** in an object can change. Temperature measures the thermal energy of an object.

☐ You should know that **electrical energy** flows through a simple electrical circuit. The circuit must be complete. Electrical energy in a current can produce thermal energy, light, sound and magnetism

☐ You should know that **light energy** can travel through various materials or through empty space. **Light** can be absorbed, transmitted, refracted (*bent*) or reflected.

☐ You should know that **sound** is caused by vibrations. Changing the rate of vibrations will change the pitch of the sound. Sound, like light, can be absorbed, transmitted, or reflected.

CHAPTER STUDY CARDS

Energy

Energy has an ability to change matter. There are many forms of energy:

★ **Electricity.** Energy of charged particles.

★ **Thermal Energy (heat energy).** Energy from motion of particles in matter. It is measured by its temperature.

★ **Light.** Can travel through space and many materials.

★ **Sound.** Energy created by vibrations carried by sound waves.

Electricity

★ Electricity flows in a complete circuit.

★ Electricity can create thermal energy, light, sound, and magnetism.

Light

★ Light travels through a single substance or space in a straight line.

★ Many materials **transmit** or absorb light.

★ **Reflection.** Some materials reflect light. The light rays bounce off these materials

★ **Refraction.** Light rays may become bent as they pass from one material into another.

CHECKING YOUR UNDERSTANDING

1. Jack arranged wire, a battery, a switch and a small light bulb in four different ways as shown below. Which bulb will light up?

PS: E
G5.4

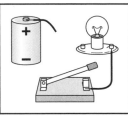
A.

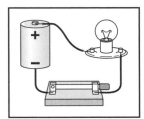
B.

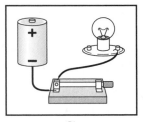
C.

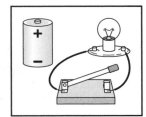
D.

HINT

To answer this question correctly, you must understand electric circuits. In a circuit such as this (wire, battery, switch, and light bulb), electricity leaves the source (battery) and travels around a pathway until it returns to the other side of the source. If you examine each diagram carefully, you will see that only one shows a complete circuit. Therefore, **Choice B** is the correct answer.

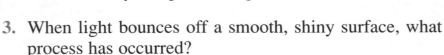

Now try answering some additional questions on your own:

2. A student goes out for a walk and sees a flash of lightning. What does this flash of lightning illustrate?

   A. Light can be reflected.
   B. Light is a form of heat.
   C. Electricity flows in a circuit.
   D. Electricity can produce light.

   PS: E
   G5.3

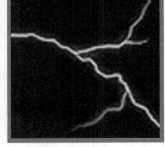

3. When light bounces off a smooth, shiny surface, what process has occurred?

   A. refraction
   B. reflection
   C. transmission
   D. photosynthesis

   PS: F
   G5.5

4. Which appliance works by turning electrical energy into thermal energy?

   A. a CD player
   B. a light switch
   C. an electric fan
   D. an electric stove

   PS: E
   G5.3

5. What does electricity traveling through a wire, battery, and light bulb illustrate?

   A. a circuit
   B. reflection
   C. photosynthesis
   D. heat conduction

   PS: E
   G5.4

6. Jared lives near an airport. Whenever a jet flies over his house, the windows of his house shake. Why does this happen?

   A. Sound waves are vibrations.
   B. Sound waves can be reflected.
   C. Electricity can create magnetism.
   D. Light energy travels faster than sound waves.

   PS: F
   G5.6

7. Which of these objects can conduct electricity?

   A. a glass dish
   B. a rubber ball
   C. a plastic hanger
   D. a copper handle

   ♦ Examine the Question
   ♦ Recall What You Know
   ♦ Apply What You Know

   PS: B
   G4.3

8. A pot of water has a temperature of 25°C. It is placed on a hot stove having a temperature of 120°C, while the temperature of the room is 20°C. Which best describes the transfer of thermal energy?

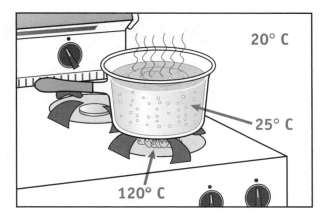

   A. stove → pot → air
   B. air → stove → pot
   C. stove → air → pot
   D. pot → air → stove

   PS: D
   G5.2

9. A student rubs two wooden sticks together very hard and fast. What effect will this rubbing have?

   A. The sticks will begin to melt.
   B. The sticks will become magnetic.
   C. The temperature of the sticks will increase.
   D. The sound of the rubbing will become lower.

   PS: D
   G4.5

10. A coil of wire is wrapped around an iron rod. When the wire is connected to a battery, the rod attracts steel paper clips. Why does the rod attract the clips?

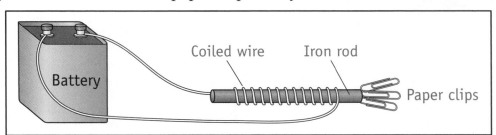

   A. The paper clips are being pulled by gravity.
   B. The paper clips have collided with the iron rod.
   C. Electricity has increased the rod's thermal energy.
   D. Electricity in the wire has made the rod magnetic.

   PS: E
   G5.3

11. The thermal energy of an object often changes.

   In your **Answer Document**, identify two ways that the thermal energy of an object might be changed. (2 points)

12. Electricity is a form of energy created by fast-moving, charged particles. Electricity has many special characteristics.

   PS: D
   G4.5

   In your **Answer Document**, identify two characteristics of electricity.

   Then, describe each characteristic. (4 points)

   PS: E
   G5.3

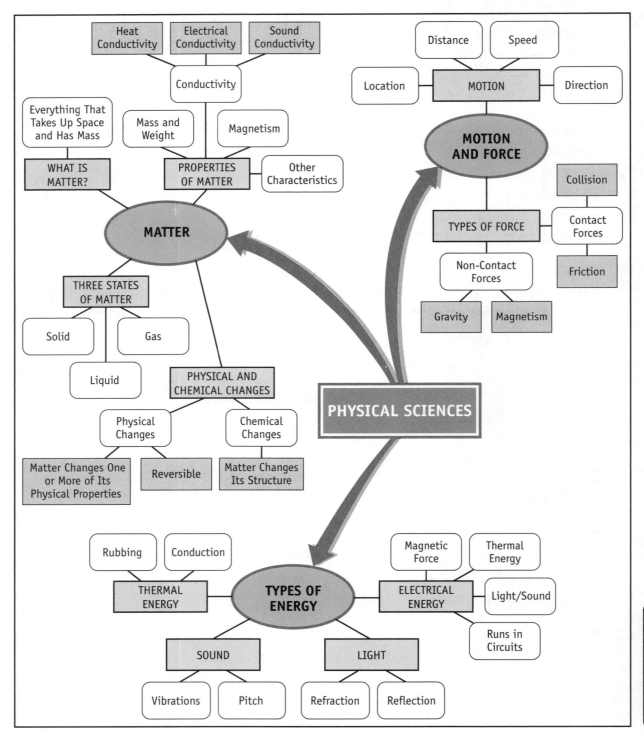

TESTING YOUR UNDERSTANDING

1. Why does blowing into a trumpet make a sound?

   A. The trumpet heats the air
   B. The trumpet reflects the air.
   C. The air in the trumpet is cooled.
   D. The trumpet causes air to vibrate.

**PS: F
G5.6**

Marble	Tennis ball	Wooden box	Small boulder
2 g	10 g	1 kg	100 kg

2. Which of these objects would require the most force to move a distance of 5 meters?

   A. marble   C. wooden box
   B. tennis ball   D. small boulder

   **PS: C
   G3.4**

3. Which of these appliances changes electrical energy into light energy?

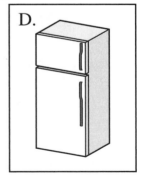

   A.   B.   C.   D.

   **PS: E
   G5.3**

4. Solids, liquids, and gases all have different properties.

   Which of the following will change shape to match its container?

   A. only solids   C. solids and liquids
   B. only liquids   D. liquids and gases

   **PS: B
   G4.4**

5. An object is placed on a desk. When a magnet is placed nearby, the object slowly moves toward the magnet. What material is the object made of?

   A. glass
   B. iron
   C. plastic
   D. rubber

   PS: B
   G4.3

6. People who live in cold climates often wear thermal clothes when they go outside. How does thermal clothing help people stay warm?

   A. Thermal clothing absorbs human perspiration.
   B. Thermal clothing slows down the loss of energy.
   C. Thermal clothing creates its own thermal energy.
   D. Thermal clothing is a good conductor of thermal energy.

   PS: D
   G5.2

7. Which of the following shows light being reflected?

   PS: F
   G5.5

A.

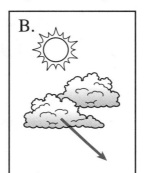

B.

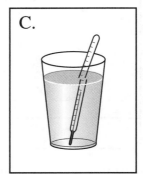

C.

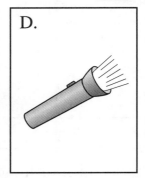

D.

8. A builder has mistakenly mixed salt with sand he needs for construction. How could the builder safely remove the salt from the mixture?

   A. freeze the mixture
   B. put the mixture in water
   C. set the mixture on fire
   D. hold a magnet by the mixture

   PS: B
   G4.3

9. A bicycle's reflector helps people to see the bicycle while it is moving at night. How does the reflector work?

   A. It refracts light rays.
   B. It bounces light back.
   C. It absorbs light shined on it.
   D. It makes light with electricity.

   PS: F
   G5.5

10. Jane's science homework asks why air is matter. What should she answer?

    A. Air is invisible.
    B. Air takes up space and has mass.
    C. Air is needed for people to breathe.
    D. Air takes the shape of its container.

    PS: B
    G4.4

11. The instrument on the right is used to test which materials conduct electricity. Which list below has three objects that conduct electricity?

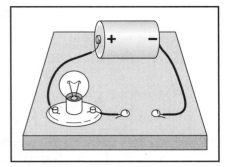

A. pencil, eraser, spoon
B. paper clip, penny, iron screw
C. cork, dollar bill, metal tweezers
D. rubber ball, plastic comb, iron nail

PS: B
G4.3

12. Which of the following would reflect rather than refract light?

A. a small lens   C. a dull black paper
B. a shiny mirror   D. a magnifying glass

PS: F
G5.5

13. Katie was in her backyard, drinking a glass of ice water on a sunny day. She placed her glass on the backyard table and went inside. When she returned, the ice in the glass had melted. What type of change took place?

A. refraction
B. photosynthesis
C. a reversible physical change
D. a reversible chemical change

◆ **Examine the Question**
◆ **Recall What You Know**
◆ **Apply What You Know**

PS: A
G4.1

14. What instrument could be used to measure the thermal energy released by these burning candles?

A. telescope
B. barometer
C. meterstick
D. thermometer

PS: D
G5.1

15. Why do many people wear wool coats in cold weather?

A. A coat prevents the loss of thermal energy.
B. A coat slows down the loss of thermal energy.
C. A coat prevents cold from entering a person's body.
D. A coat slows down cold from entering a person's body.

PS: D
G5.2

16. A student is playing a guitar. Some of the strings vibrate more quickly than other strings. What effect does this have on the sound?

A. The faster vibrating strings make louder sounds.
B. The faster vibrating strings make lower sounds.
C. The faster vibrating strings make softer sounds.
D. The faster vibrating strings make higher sounds.

PS: F
G5.7

17. What does the temperature of a beaker of water measure?

A. its weight
B. its thermal energy
C. its magnetic force
D. its transmission of light

PS: D G5.1

18. A student places an eraser on a desktop with a pencil and a paper clip.

In your **Answer Document**, give two facts that describe the location of the eraser. (2 points)

PS: C G3.1

19. There are three different states that matter can take.

In your **Answer Document**, identify two of these states of matter.

Then, describe one characteristic of each state. (4 points)

PS: B G4.4

CHECKLIST OF SCIENCE BENCHMARKS

Directions. Now that you have completed this unit, place a check (✔) next to those benchmarks you understand. If you are having trouble recalling information about any of these benchmarks, review the lesson indicated in the brackets.

PHYSICAL SCIENCES

☐ You should be able to compare the characteristics of simple physical and chemical changes. [**Lesson 11**]

☐ You should be able to identify and describe the physical properties of matter in its various states. [**Lesson 11**]

☐ You should be able to describe the forces that directly affect objects and their motion. [**Lesson 12**]

☐ You should be able to summarize the ways that changes in temperature can be produced and thermal energy transferred. [**Lesson 13**]

☐ You should be able to trace how electrical energy flows through a simple electrical circuit and describe how the electrical energy in a circuit can produce thermal energy, light, sound and magnetic force. [**Lesson 13**]

☐ You should be able to describe the properties of light and sound energy. [**Lesson 13**]

EARTH AND SPACE SCIENCES

A view of Earth from Apollo 17.

In this unit, you will review what you need to know about the Earth and Space Sciences. First, you will learn about the stars. You will also learn that our sun is a star. Next, you will learn about the planets in our solar system, including Earth. You will also learn how Earth rotates on its axis and orbits the sun.

Then you will learn more about the forces that are continually shaping Earth's surface. Lastly, you will learn about Earth's resources and weather.

LESSON 14: EARTH'S PLACE IN THE UNIVERSE
In this lesson, you'll learn about Earth's position in space — how it rotates as it revolves around the sun. You will also learn about the sun, planets, and more distant stars.

LESSON 15: EARTH'S CHANGING SURFACE
In this lesson, you will learn about the processes shaping Earth's land surface areas. You will learn how wind, water, ice, and plant growth weather and erode rocks and soil, and how other processes build up new landforms.

LESSON 16: EARTH'S RESOURCES
In this lesson, you will learn about Earth's resources — especially its rocks and soil. You will also learn why Earth's renewable and non-renewable resources need to be protected.

LESSON 17: WEATHER
In this lesson, you will learn about the weather, including barometric pressure, the effects of water in the air, types of clouds, and different weather patterns.

LESSON 14

EARTH'S PLACE IN THE UNIVERSE

In this lesson, you will learn about the solar system. You will also learn how Earth rotates on its axis while orbiting the sun.

— MAJOR IDEAS —

A. Earth spins or **rotates** on its axis. This spinning causes us to have day and night. Earth is **tilted** on its **axis** as it orbits around the sun. This explains why the seasons of the year change from spring and summer to fall and winter.

B. The **solar system** consists of the sun and eight planets — *Mercury, Venus, Earth, Mars, Jupiter, Saturn, Uranus,* and *Neptune.*

C. The **moon** orbits Earth.

D. The **sun** is a star. It is the major source of energy for Earth.

E. **Stars** are enormous balls of superheated gases, like the sun. Because other stars are so far away, they appear to be points of light in the sky.

OUR SOLAR SYSTEM

Most of the universe is empty space. In this empty space are **galaxies** — groups of stars. Our sun is one of these stars. The sun is the center of our solar system. Our planet, Earth, is also part of this vast system.

STARS

Stars are enormous balls of superheated gases. The nearest star to Earth — after the sun — is 40 trillion kilometers (*25 trillion miles*) away.

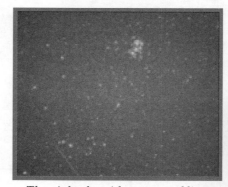

The night sky with stars sparkling.

164

How Stars Appear in the Sky. Because other stars are so far away, they look to us like points of light in the night sky. Actually, each star is gigantic in size. The center of a star is extremely hot and dense. The star's internal pressure causes tiny particles of matter to join together. This process releases tremendous amounts of energy, which moves outward from the star's center. This energy finally radiates from the star's surface across space. Stars can keep producing energy for billions of years.

THE SUN

The **sun** is a star. It appears to be the largest star in the sky because it is the closest star to Earth. The sun's reactions produce enormous amounts of energy, giving off light that travels through space. The sun is the source of most energy on Earth and the rest of the solar system. It is the source of our heat and allows living things on our planet to survive. The sun is the main influence on Earth's climate and weather.

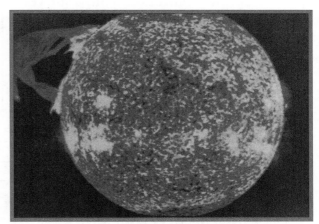

The sun is the closest star to Earth.

THE PLANETS

Planets are objects of rock, metal, ice and gas that orbit the sun. They do not give off their own light as the sun does. There are eight planets in our solar system. They range in size from small rocky planets to huge gas giants with rings. The planets orbit the

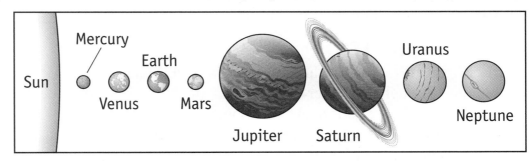

sun. In the order of their distance from the sun, the eight planets are *Mercury*, *Venus*, *Earth*, *Mars*, *Jupiter*, *Saturn*, *Uranus*, and *Neptune*. The largest planet is Jupiter. It is so large that all the other planets could fit inside it. Scientists once believed there was a ninth planet, Pluto. Because of its small size, scientists now consider Pluto a dwarf planet. There are other dwarf planets, such as Eris, which is slightly larger than Pluto. Our solar system also contains asteroids and comets.

APPLYING WHAT YOU HAVE LEARNED

✦ This table shows that Earth is 149.6 million kilometers from the sun, or 93 million miles away. Suppose there was an imaginary road from Earth to the sun. If you were in a car traveling 100 km an hour, it would take you just over 170 years to reach the sun. If it took 170 years to drive from Earth to the sun, about how many years would it take to drive from Jupiter to the sun at the same speed?

DISTANCE OF PLANETS FROM THE SUN

Planet	Distance from the Sun
Mercury	57.9 million km
Venus	108.2 million km
Earth	149.6 million km
Mars	227.9 million km
Jupiter	778.3 million km
Saturn	1,427.0 million km
Uranus	2,871.0 million km
Neptune	4,497.1 million km

Your answer: _____

THE MOVEMENT OF THE EARTH AND MOON

Earth is a very unique planet. Scientists believe it is the only planet in our solar system where liquid water is present. In fact, three-quarters of Earth's surface is covered by water. In addition, our planet is the only one wrapped in a thin blanket of air. Earth actually moves through space in two different ways at the same time: it **rotates** on its axis, and it **orbits** the sun.

EARTH'S ROTATION

The Earth **rotates**, or spins, around its **axis** — an imaginary line running through the center of Earth from the North Pole to the South Pole. This rotation takes 24 hours, causing day and night on Earth. Night occurs on those parts of Earth that are away from the sun's rays. The sun, however, is not circling Earth. It just appears that way because of Earth's rotation. The sun appears to rise in the east and set (*go down*) in the west.

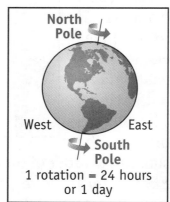

North Pole

West   East

South Pole

1 rotation = 24 hours or 1 day

EARTH ORBITS THE SUN

Earth also **orbits** or **revolves** around the sun while it rotates on its axis. The shape of Earth's orbit is **elliptical** — like an oval. It takes just over 365 days (*one year*) for Earth to complete one **orbit** around the sun.

Earth tilts on its **axis** as it orbits the sun. Because of this tilt, the sun's rays hit the Northern Hemisphere more directly in summer than in winter. During summer, days are longer and temperatures are warmer. Because the sun's rays are direct, shadows are shorter. When it is sum-

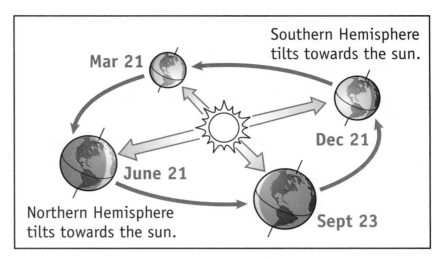

mer in the Northern Hemisphere, it is winter in the Southern Hemisphere. This is because the Southern Hemisphere tilts away from the sun and receives less direct sunlight. In winter, days are shorter, temperatures are cooler, and shadows are longer.

APPLYING WHAT YOU HAVE LEARNED

1. Scientists see Earth as a unique planet. What are some of the characteristics that make Earth unique? _____

   _____

2. "The planet Earth moves in two different ways at the same time." Explain this statement. _____

   _____

MOVEMENT OF THE MOON

Earth has one natural satellite, the **moon**. The moon orbits Earth every $29\frac{1}{2}$ days. The moon is about one-quarter the size of Earth. Because of its smaller size, the moon's gravity is only a fraction of Earth's gravity.

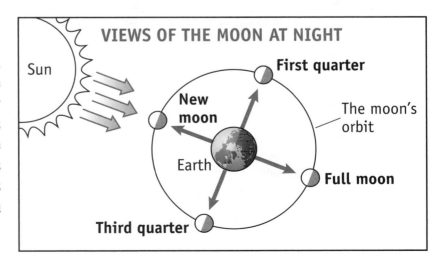

Although we see the moon in the night sky, it does not create its own light. Moonlight is light that is reflected from the sun. The size of the bright portion of the moon changes throughout the month as the moon moves around Earth.

APPLYING WHAT YOU HAVE LEARNED

✦ Explain how movements of Earth and the moon are alike and different.

_____

WHAT YOU SHOULD KNOW

☐ You should know that **stars** are enormous balls of superheated gases. The **sun** is a star. It is the major source of energy for Earth.

☐ You should know that the **solar system** consists of the sun and eight planets — _Mercury, Venus, Earth, Mars, Jupiter, Saturn, Uranus,_ and _Neptune._

☐ You should know that Earth **rotates** on its axis. This spinning causes us to have day and night. Earth is **tilted** on its axis as it orbits around the sun. This explains why seasons change from spring and summer to fall and winter.

☐ You should know that Earth is wrapped in a blanket of air, and that three-fourths of Earth's surface is covered by water.

☐ You should know that the moon orbits Earth.

CHAPTER STUDY CARDS

The Solar System

★ **Stars.** Stars are enormous balls of super-heated gases.

★ **The sun.** The sun is a star. It is the largest body in the solar system. The sun is the source of most of our energy.

★ **Planets.** There are eight planets. Each planet orbits the sun. The planets, in their order from the sun, are Mercury, Venus, Earth, Mars, Jupiter, Saturn, Uranus, and Neptune.

Earth and Moon

★ **Moon.** The moon is a satellite of Earth that orbits Earth.

★ **Earth.** Earth rotates on its axis every 24 hours, creating night and day.

• Earth completely orbits around the sun once every year. Earth's orbit around the sun is elliptical (oval).

• Earth's tilt on its axis gives us our four seasons: spring, summer, fall, winter.

• The sun appears to rise in the east and set (go down) in the west.

CHECKING YOUR UNDERSTANDING

1. Which correctly describes the movement of bodies in space?

   A. Earth orbits the moon.
   B. The moon orbits Earth.
   C. The sun orbits the moon.
   D. Earth and the moon orbit each other.

   ES: A
   G5.2

   HINT To answer this question, you should know that Earth is one of several planets that orbit the sun, and that the moon orbits Earth. Only **Choice B** correctly describes these movements.

   Now try answering some additional questions on your own:

2. Which illustration correctly depicts the movement of Earth, the sun and the moon?

   ES: A
   G5.2

   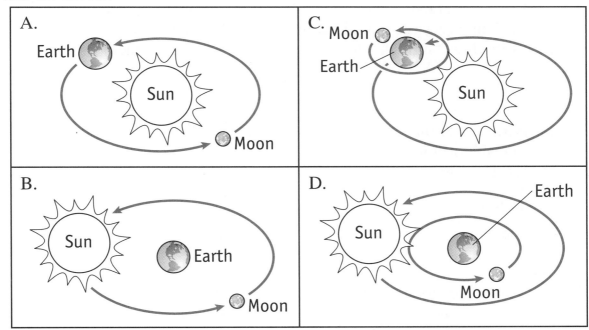

3. What is a result of the rotation of Earth?

   A. four seasons
   B. day and night
   C. the calendar year
   D. the moon's changing size

   ♦ **Examine the Question**
   ♦ **Recall What You Know**
   ♦ **Apply What You Know**

   ES: A
   G5.1

4. Which best explains why the amount of daylight increases each day in Ohio as the summer months approach?

   A. Earth is moving closer to the sun.
   B. The rotation of Earth is slowing down.
   C. The moon is reflecting more light from the sun.
   D. The Northern Hemisphere is tilting more toward the sun.

   ES: A
   G5.3

5. Which best describes Earth's orbit around the sun?

   A. zig-zag   C. elliptical
   B. circular   D. rectangular

   ES: A
   G5.3

6. What determines the length of one day on Earth?

   A. The time it takes Earth to circle the sun.
   B. The time it takes the moon to circle Earth.
   C. The time it takes Earth to rotate on its axis.
   D. The time it takes the sun to spin around its axis.

   ES: A
   G5.1

7. Why is the Northern Hemisphere warmer in summer than in winter?

   A. Earth is moving more quickly in its orbit around the sun.
   B. Less sunlight shines on the Northern Hemisphere in summer.
   C. The sun gives off more heat in the summer than in the winter.
   D. More direct sunlight shines on the Northern Hemisphere in summer.

   ES: A
   G5.3

8. Why does the sun appear to rise and fall in the sky each day?

   A. The sun rotates.
   B. Earth orbits the sun.
   C. The sun orbits Earth.
   D. Earth rotates on its axis.

   ◆ Examine the Question
   ◆ Recall What You Know
   ◆ Apply What You Know

   ES: A
   G5.1

9. About how long does it take Earth to complete one orbit around the sun?

   A. one day   C. one year
   B. one month   D. two years

   ES: A
   G5.3

10. Planet Earth is unique among the planets in our solar system.

    In your **Answer Document**, identify two characteristics that make Earth unique. (2 points)

    ES: A
    G5.3

11. Earth moves in two directions at the same time.

    In your **Answer Document**, explain how Earth moves in two ways at the same time. (2 points)

    ES: A
    G5.2

EARTH'S CHANGING SURFACE

When you look at Earth's mountains, oceans, forests, and deserts, you may think these surface features have remained unchanged for millions of years. Nothing could be further from the truth. In fact, Earth's surface features are in a continual state of change. In this lesson, you will learn about some of the processes (*events leading to change*) that shape Earth's surface.

— MAJOR IDEAS —

A. Freezing, thawing, and plant growth cause the **weathering** of rocks.

B. Wind, water, and ice can **erode** rock and **deposit** their particles in other areas.

C. Some processes that shape Earth's surface — like weathering, erosion, deposition, and mountain building — are slow.

D. Other processes — like volcanic eruptions, earthquakes and landslides — bring about rapid changes to Earth's surface.

WEATHERING

The wearing down of rocks on Earth's surface by wind, water, ice and plants is known as **weathering**. For example, cool nights and hot days can cause rocks to crack. Water may seep into the cracks in the rocks. The water expands if it freezes, increasing the cracks and breaking the rock apart. It melts and freezes again, repeating the process. Rain and running water will also

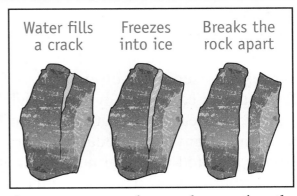

Water fills a crack Freezes into ice Breaks the rock apart

break down rock into smaller particles. Running water smoothes rock, creating the round stones and pebbles found in streams and rivers.

Plants also wedge their roots into the cracks of rocks. As they grow, the plants spread these cracks wider. Eventually, the rocks break apart.

APPLYING WHAT YOU HAVE LEARNED

1. Describe different kinds of weathering that can wear down rocks. _____
   _____
   _____

2. Why do you think this process is called "weathering"? _____
   _____

EROSION

Erosion is the process by which soil and rock are broken down and moved away. Once rock is broken down into smaller particles, then wind, running water, or ice often carries these particles to a new location.

WIND

If you've ever been to a beach on a windy day, you can understand the power of sand blown by the wind. Over time, particles in the wind can wear down rock. Wind can also blow away dry soil.

WATER

Running water in streams and rivers can also break up rock and soil. The water then carries this sediment elsewhere. A fast-flowing river can gradually cut through layers of rock, creating a beautiful **canyon**, like the Grand Canyon. To reduce soil erosion from water and wind, farmers often plant trees, crops or grass. The roots of the plants help hold the soil together against the wind

Erosion at the Grand Canyon (USGS)

and rain. Farmers also plow the land on hills in horizontal rows to prevent fast-running rain water from falling straight downwards and washing away topsoil.

GLACIERS

Ice can also cause erosion. **Glaciers** are rivers of ice. They form in places where there are very cold winters and cool summers. The snow that falls in the winter does not melt during the summer. Instead, it turns to ice. New snow then falls on

top of this ice. As the layers of snow build up, the weight of the snow increases. This weight pushes on the ice and snow below, creating thick, dense sheets of ice.

How Glaciers Cause Erosion. Glaciers move quite slowly. As they move, they scrape Earth's surface. Glaciers can pick up loose rock in their way, dig holes, wear down mountains, and move rocks and soil. Often, moving glaciers will carve valleys through mountains. The rocks and boulders carried by the glaciers scrape the surface. A glacier can move millions of tons of material. During the last Ice Age, part of Ohio was covered by glaciers.

APPLYING WHAT YOU HAVE LEARNED

1. What is a glacier? _____

   _____

2. How are glaciers formed? _____

   _____

3. How do glaciers cause erosion? _____

   _____

DEPOSITION

Deposition is the process by which the rocks, soil, and other sediment moved by erosion are left or deposited in a new place. This creates typical landforms, such as dunes, deltas, and glacial moraines.

DUNES

Dunes are hills of sand. They are most common in dry inland areas where lakes or seas once existed. Winds blow the sand, creating hills with crests or ridges. Dunes are also found in coastal regions, where winds blow the sand. Some dunes in California have been known to advance 50 feet a year. Such deposition can create a serious threat to nearby farms and highways.

Medano Creek and Cleveland Peak Dunes

DELTAS

A **delta** is an area that forms where a river flows into an ocean, sea, or lake. The river carries soil and other sediment. Once the river reaches the ocean or sea, its current stops. The sediment it is carrying is deposited at its mouth. It gradually builds up, extending the land in a typical triangle shape. For example, over the past 5,000 years, the Mississippi River has caused parts of the U.S. coastline to advance 50 miles into the Gulf of Mexico.

The Mississippi Delta seen from space.

GLACIAL MORAINES

A **glacial moraine** is material moved by a glacier and then left when the glacier retreats. It consists of the rocks, gravel, sand and soil scraped by the glacier from Earth's surface. The material may be carried by the glacier on its surface or underneath. Moraines typically appear as hills.

Moraines have formed on both sides of the glacier.

APPLYING WHAT YOU HAVE LEARNED

Complete the following chart:

Deposition	How It Is Formed	What It Looks Like
Dune		
Delta		
Moraine		

MOUNTAIN BUILDING

Earth's surface is shaped by both slow and rapid processes. Weathering, erosion, and deposition are slow processes. They shape Earth's surface gradually. Another important process that shapes Earth's landforms gradually is **mountain-building**.

Sometimes sections of Earth's crust (*its outermost layer*) slowly squeeze together over millions of years. When this happens, Earth's crust may start to fold. This folding pushes some of Earth's crust upwards. This process can form **mountains**. Many of the great mountain ranges on Earth were created by this slow folding of Earth's crust.

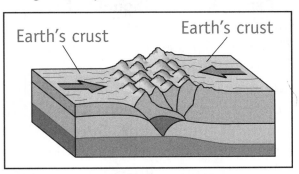

The movement of Earth's crust builds mountains.

RAPID PROCESSES SHAPING EARTH

Earth's surface sometimes changes more suddenly. Volcanic eruptions, earthquakes and landslides can rapidly change the face of Earth.

VOLCANIC ERUPTIONS

A **volcano** is an opening in Earth's surface that lets out molten rock and gases. Volcanoes often occur at the edges of huge plates of Earth's crust that shift back and forth over time. Sometimes, one of these plates will sink into Earth. Then it creates molten rock known as **magma**. A volcano **erupts** as magma escapes through a hole in Earth's crust. Once this molten rock comes to the surface, it becomes known as **lava**. As lava builds up on the ground around the volcano, it gives the volcano a typical round shape, like an upside-down cone. Many islands and mountains have been formed by volcanoes. For example, the Hawaiian Islands are actually the tops of volcanoes in the Pacific Ocean.

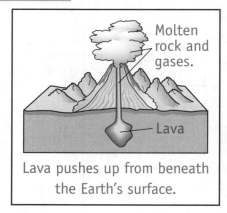

Lava pushes up from beneath the Earth's surface.

EARTHQUAKES

Another process leading to rapid change occurs when movements of Earth's crust create pressure and stress in the surrounding rock. Eventually, the rock releases the energy created by this stress in an **earthquake**. The rock vibrates to release energy. This energy passes through Earth to the surface in a series of **seismic waves**. An earthquake may tear down an area or help build it up. Lava may come through cracks created by the earthquake, creating a mountain range. An earthquake may shift Earth's crust, lowering some areas and raising others up.

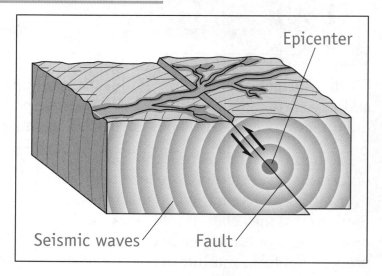

LANDSLIDES

A landslide is another process that can change Earth's surface rapidly. A **landslide** occurs when rocks, soil, and other materials drop down a steep slope, mountain or cliff. Gravity pulls these materials downward. The slope might have been weakened by erosion, heavy rain, melting snow, earthquakes, or groundwater. An **avalanche** occurs when snow, ice, and rock fall quickly down the side of a mountain.

APPLYING WHAT YOU HAVE LEARNED

Explain how each of these processes affects Earth's surface.

Process	What Is It?	Effects On Earth's Surface
Mountain-building		
Volcanic Eruptions		
Earthquakes		
Landslides		

WHAT YOU SHOULD KNOW

☐ You should know that freezing, thawing, and plant growth can cause the **weathering** of rocks.

☐ You should know that wind, water, and ice **erode** rock and soil in some areas and **deposit** these particles in other areas.

☐ You should know that some processes that shape Earth's surface — like **weathering**, **erosion**, **deposition,** and **mountain building** — are slow.

☐ You should know that other processes — like **volcanic eruptions**, **earthquakes** and **landsides** — bring about rapid changes in Earth's surface.

CHAPTER STUDY CARDS

Rapid Processes

★ **Volcanoes.** Molten rock (**lava**) from underground breaks through Earth's surface and hardens when cool.

★ **Earthquakes.** Part of Earth's crust vibrates to release stress. This energy passes to Earth's surface in a series of **seismic waves**.

★ **Landslides.** Rock, soil, snow, ice and other materials fall off a weakened slope, cliff, or mountainside.

Gradual Processes

★ **Weathering.** Wind, water, ice and plants wear down rocks. Water freezes, thaws, and freezes again to break up the rock.

★ **Erosion.** Soil and rock are broken down and carried away by wind and water.

★ **Glaciers.** Giant sheets of ice that move slowly, scraping Earth's surface.

★ **Folding of Earth's Crust.** Builds mountains.

★ **Deposition.** Particles are deposited in a new place: **dunes**, **deltas**, and **glacial moraines**.

CHECKING YOUR UNDERSTANDING

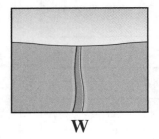 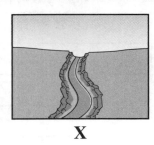 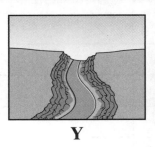 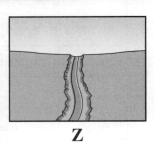

| W | X | Y | Z |

1. These diagrams show how a river can cause the erosion of land. Which is the correct order of letters that show how the river has aged from youngest to oldest?

    A. Z → W → X → Y
    B. W → Z → X → Y
    C. X → W → T → Z
    D. Y → W → X → Z

    ◆ Examine the Question
    ◆ Recall What You Know
    ◆ Apply What You Know

    ES: B
    G4.8

HINT To answer this question, you need to know how water erosion gradually grinds down rock and moves rock particles away. As the river continues to flow, it makes the gap wider. The correct answer goes from the smallest (W) to the largest (Y) — **Choice B**.

Now try answering some additional questions on your own:

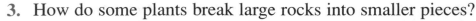

2. What is the most common cause of earthquakes?
    A. the sinking of the ocean floor
    B. movements of the Earth's crust
    C. unequal heating of the atmosphere
    D. giant waves caused by the pull of the moon

    ES: B
    G4.10

3. How do some plants break large rocks into smaller pieces?
    A. Plant stems surround and squeeze rocks until they break.
    B. Plant roots grow into cracks in rocks, forcing them apart.
    C. Plant leaves protect the rocks from extreme temperatures.
    D. Plants fall into rocks and release their powerful chemicals.

    ES: B
    G4.9

4. A farmer wishes to reduce the erosion of topsoil caused by heavy spring rains on his farm. How should he plow his fields?

ES: B
G4.8

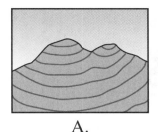
A.

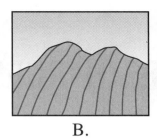
B.

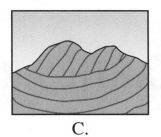
C.

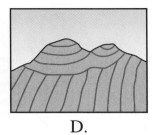
D.

5. How does freezing water cause rocks to weather?

A. It expands their cracks.
B. It makes them colder.
C. It holds them in place.
D. It smooths their edges.

ES: B
G4.9

6. Theresa is visiting the desert for the first time. She is surprised to see hills of sand formed by the action of the wind. What do scientists call these hills?

A. deltas
B. dunes
C. moraines
D. volcanoes

◆ Examine the Question
◆ Recall What You Know
◆ Apply What You Know

ES: B
G4.8

7. A science class built the models shown to the right to conduct an experiment. Students poured the same amount of water over both models. They observed that in Model A, most of the soil emptied out of the container. In Model B, only a small amount of soil emptied out of the container. What were the students studying in their experiment?

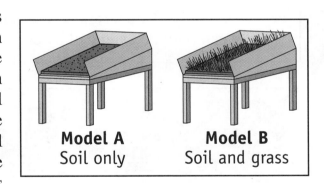

Model A
Soil only

Model B
Soil and grass

A. the rock cycle
B. the impact of volcanoes
C. the force of earthquakes
D. the effects of water erosion

ES: B
G4.10

8. Rocks are constantly undergoing change from weathering.

In your **Answer Document**, describe two forces that cause rocks to weather. (2 points)

ES: B
G4.9

LESSON 16

EARTH'S RESOURCES

In this lesson, you will learn about some of the resources that make up Earth — including rocks, soil, water and air. You will also learn why it is important to conserve both renewable and non-renewable resources.

— MAJOR IDEAS —

A. **Rocks** have distinct properties, such as color, layering, and texture. These properties help scientists to classify rocks.

B. **Soil** is made up of small pieces of rock and decomposed pieces of plants and animals.

C. Soils have different properties, including color, texture, the ability to hold water, and the ability to support life.

D. Earth's **renewable resources** — such as trees, fresh water, air, and wildlife — can only be maintained with careful use.

E. The supply of **non-renewable resources**, like coal, is limited. Their supply can only be extended by conservation and recycling.

EARTH'S RESOURCES

Earth provides many different natural resources that people need and use. Two of the most important resources are rocks and soil.

ROCKS

Rocks should be quite familiar to you. You can see them all around you in mountains, canyons and riverbeds. A **rock** is any natural solid found on Earth's surface or below it.

180

Rocks are made of one or more **minerals**. Many minerals form geometric shapes known as **crystals**. Most of Earth consists of hard, dense rock or molten material made from rock. As you go deeper into Earth, this rock becomes hotter.

TYPES OF ROCKS

There are many different kinds of rocks. Scientists classify rocks in a variety of ways — such as by their color, size, and how the rock was formed. One common way to identify a rock is by the **mineral crystals** it contains. Many rocks have crystals of quartz, mica, or feldspar. Scientists also sometimes identify a rock by its **hardness** or **texture** — how the rock feels when you touch it. Lastly, many rocks have different **layers**. The layers of a rock help show how it was formed. In general, there are three types of rocks:

★ Some rocks (*igneous*) have formed from magma or lava that has cooled. These rocks are usually heavy and they often have speckles, mineral crystals, grains, or veins. When the magma took a long time to cool, the mineral crystals in the rock are larger. Granite is an example of this kind of rock.

A rock formed from cooled magma.

★ Other rocks (*sedimentary*) build up in layers. When rock erodes, its particles are moved to a different place. Pieces of weathered rock and other materials become deposited on top of each other. This sediment may be pressed into layers. The top layers push down on the lower layers with their weight. Eventually, these layers of sediment turn into rock. These rocks are usually lighter in color and weight than rocks from cooled magma. Their layers are often vis-

A rock formed from layers of sediment.

ible. Fossils are found in this kind of rock. Examples of rock made by sediment are limestone and sandstone.

★ A third type of rock (*metamorphic*) is formed by rocks that have changed into another kind of rock. These rocks may have moved back under Earth's surface. There, over millions of years, heat and pressure change this rock into a new rock, such as marble or slate.

A rock formed by heat and pressure.

All rocks can be classified as one of these three main types based on the way they were formed. Smaller rocks come from the breakdown of these larger rocks by weathering and erosion.

APPLYING WHAT YOU HAVE LEARNED

Complete the following chart.

How Formed	Characteristics	Example
Some rocks form from cooled magma.		
Other rocks form from layers of sediment pressed together.		
A third type of rock is formed by heat and pressure under Earth's surface.		

SOIL

Some people just call it "dirt," but **soil** is actually essential to life on Earth. Soil is needed for growing crops and to feed humans and animals. It is important for plants because soil stores nutrients they need and provides them with support.

APPLYING WHAT YOU HAVE LEARNED

✦ What role does soil play in helping to maintain life on Earth? _____

_____

_____

WHAT IS SOIL?

Weathering breaks down rocks. The material left from the rocks mixes with decaying plants and animals and their waste products. **Soil** is therefore a mixture of several materials, including sand, clay, rock, water, fungi, bacteria, and decayed plants and animal material (*humus*). *Sand* consists of small stone particles in the soil. *Silt* feels smooth and powdery, while *clay* is the smallest type of particle found in soil. It can take hundreds of years to form one cubic inch of soil.

TYPES OF SOIL

There are different types of soils, based on the mix of materials they contain. Each type of soil has its own texture or feel, its own ability to hold water, and its own ability to support life. Soil is able to store water and nutrients used by plants because of its clay and dead plant and animal matter. Soils with a large amount of clay and decayed material hold more water than sandy soils. **Soil texture** is based on how large the pieces of clay and other particles in the soil are, and how much decayed material there is. Soils also contain different chemicals, like salts. These chemicals affect the ability of soil to support life. The type of soil differs from place to place. This explains why some areas are better for growing crops than other areas. Even in a single place, different soils usually exist in layers. The top layer, known as the **topsoil**, has more decayed matter from plants and animals. This has nutrients. Farmers can change the soil by adding different materials and chemicals. For example, farmers may add more dead plant and animal material (*humus*) or special chemical fertilizers to help crops grow.

If a particle of **sand** were the size of a basketball, then **silt** would be the size of a baseball, and **clay** would be the size of a golf ball.

PROTECTING THE SOIL

Soil is a very important resource. To preserve the soil against erosion, farmers take measures of **soil conservation**. You learned about some of these measures in the last lesson. Farmers plant crops to hold the soil together against the wind or heavy rains. Farmers also plow across hills and slopes instead of up and down to reduce soil erosion from running water.

Soil erosion from the flow of water.

APPLYING WHAT YOU HAVE LEARNED

◆ List two natural forces that cause **soil erosion** and some **measures** that farmers can take to reduce erosion. _____

_____

CONSERVING EARTH'S RESOURCES

Rocks and soil are just two of Earth's resources. Other natural resources include fresh water, air, wildlife, trees, fossil fuels and minerals. These resources can be classified into two main groups:

| Trees/Wildlife |
| Fresh Water |
| Air |

Renewable Resources

Non-renewable Resources

| Minerals |
| Fossil Fuels |

RENEWABLE RESOURCES

A **renewable resource** is something that can be replaced by natural processes like growth.

Forests are a renewable resource.

TREES AND WILDLIFE
For example, **trees** can be replaced because new trees can be regrown. If a tree is planted each time one is cut down, the resource will eventually grow back and be renewed. It is also much the same with wild animals. **Wildlife** can reproduce. However, enough time must be allowed for these resources to renew themselves.

FRESH WATER
Fresh water is another renewable resource. Water can be recycled. Waste water can be filtered to remove impurities and be used again. In the next lesson, you will learn how Earth itself recycles water. However, when rivers and lakes become polluted with liquid or solid wastes, the use of this precious resource is threatened.

AIR
Air is a third example of a renewable resource. Our Earth is surrounded by a blanket of air that we call the **atmosphere**. It reaches almost 350 miles above the surface of Earth. Our atmosphere is a mixture of different gases, especially nitrogen and oxygen. Humans and other animals need this oxygen to live. Plants take carbon dioxide from the atmosphere, which they need for photosynthesis, and emit oxygen.

The atmosphere also plays a role in recycling water. The atmosphere protects living things from the harmful effects of ultraviolet radiation given off by the sun.

A forest in California, destroyed by too much pollution in the air.

Pollution from cars and factories now threatens our atmosphere. The burning of oil, coal, and gasoline has increased levels of carbon dioxide in the atmosphere. At the same time, the cutting down of trees has reduced the production of oxygen. Carbon dioxide in the atmosphere acts as a blanket, trapping in heat. This has contributed to **global warming** — higher temperatures across the globe.

APPLYING WHAT YOU HAVE LEARNED

✦ List three ways in which life on Earth depends on the atmosphere:

1. _____

2. _____

3. _____

NON-RENEWABLE RESOURCES

A **non-renewable resource** is one that was formed by Earth over millions or even billions of years and cannot be replaced or renewed. For example, oil, coal, copper and other minerals are non-renewable resources.

MINERALS

Rocks found above and below Earth's surface contain many valuable minerals. The supply of minerals like gold and iron ore are limited. Because such minerals cannot be replanted or replaced after they are taken from the ground, they are called **non-renewable resources**.

A raw gold specimen

FOSSIL FUELS

Fossil fuels — like coal, oil, and natural gas — are very special resources. They can be burned to release large amounts of energy. We burn fossil fuels to run our car engines, heat our homes, power our machinery, and create electricity. Fossil fuels actually come from the remains of ancient living things.

★ **Coal** is a brown or black rock formed from plants in ancient forests and swamps as long as 400 million years ago. After the plants died, they decayed. Over millions of years, heat and pressure changed their remains into coal. Today, we burn coal for electricity and heat. When burned, coal releases the energy stored by plants from the sun many millions of years ago.

★ **Oil and natural gas** are also fossil fuels. They were formed by very tiny plants and animals in the ocean. These living things stored energy, originally taken from the sun through photosynthesis. When they died, they settled or fell to the ocean floor, where mud and sediment covered them. Over millions of years, heat and pressure changed their soft bodies into liquid oil and natural gas.

OIL (PETROLEUM) AND NATURAL GAS FORMATION

300–400 million years ago

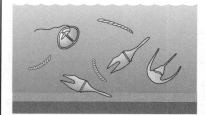

Tiny sea plants and animals died and were buried on the ocean floor. Over time, they were covered by layers of silt and sand.

50–100 million years ago

Over millions of years, the remains were buried deeper and deeper. The enormous heat and pressure turned them into oil and gas.

Today

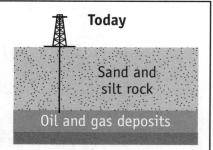

Today, we drill down through layers of sand, silt, and rock to reach the rock formations that contain oil and gas deposits.

It takes millions of years for fossil fuels like coal and oil to form. They can only be burned once. For this reason, they are considered important **non-renewable** resources. Although non-renewable resources cannot be replaced, the use of these resources can be extended by careful use without waste. **Conservation** and **recycling** are just two ways to help extend these resources:

★ When we **conserve**, we use less of a resource. Newer car engines burn less gasoline than older models.

★ When we **recycle**, we reuse a resource. We may collect and reuse old newspapers to make paper instead of cutting down new trees.

APPLYING WHAT YOU HAVE LEARNED

1. Why are scientists concerned about the rate humans are using up fossil fuels?

   _____

   _____

2. What steps can be taken to help the world's supply of fossil fuels last longer?

   _____

   _____

3. Classify each of the following resources by checking the correct box:

	Renewable	Non-Renewable		Renewable	Non-Renewable
Gold	☐	☐	Oak trees	☐	☐
Iron ore	☐	☐	Rubber plants	☐	☐
Coal	☐	☐	Wild salmon	☐	☐
Fresh water	☐	☐	Petroleum	☐	☐

WHAT YOU SHOULD KNOW

■ You should know that **rocks** have distinct properties, such as color, layering, and texture. These properties help scientists to classify rocks.

■ You should know that rocks can be classified based on how they were formed.

■ You should know that **soil** is made up of small pieces of rock and decomposed pieces of plants and animals.

■ You should know that soils have different properties, including color, texture, the ability to hold water, and the ability to support life.

■ You should know that Earth's **renewable resources** — such as fresh water, air, wildlife, and trees — can only be maintained with careful use.

■ You should know that the supply of **non-renewable resources**, like coal, is limited. Their supply can only be extended by conservation and recycling.

CHAPTER STUDY CARDS

Earth's Resources

Earth has many different types of materials:

★ **Rock.** A solid found on Earth's crust, made of minerals. Rocks are classified by color, textures, layers, and formation.

★ **Soil.** Material from ground rock and decayed plant and animal material. Soils differ based on their mix of materials. Some hold water or support plants better than others. Farmers often take steps to conserve soils.

Types of Natural Resources

★ **Renewable Resources.** Resources that can be replaced in a shorter time span, such as:
   • **Trees**   • **Air**
   • **Fresh water**   • **Wildlife.**
The air in the atmosphere protects all life, but is threatened by pollution.

★ **Non-renewable Resources.** Resources formed by Earth over millions of years, such as:
   • **Oil**   • **Coal**   • **Minerals**
These cannot be replaced but their use can be extended by conservation and recycling.

CHECKING YOUR UNDERSTANDING

1. This rock was brought to school. The class found fossils of water plants and shells in the rock. What does this tell us about the rock?

   A. The rock is a type of marble.
   B. The rock is heavier than most rocks.
   C. The rock was formed from cooled magma.
   D. The rock was once at the bottom of the sea.

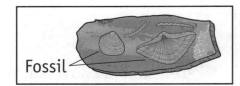
Fossil

ES: C
G3.2

 HINT   To answer this question, you need to know how different rocks are formed. This rock has seashells and fossils in it. It is these materials that make up part of the rock. This rock must have formed from pressed sediment at the bottom of the sea. Thus, the correct answer is **Choice D.**

Now try answering some additional questions on your own:

2. Which of these resources for home building is renewable?
   A. oil   C. lumber
   B. copper   D. aluminum

ES: C
G5.6

3. A farmer sees an area of land he believes will be good for farming because of its rich topsoil. Which process contributed most to forming this topsoil?

   A. sand storms
   B. ocean currents
   C. decaying of plant life
   D. folding of Earth's crust

   ♦ Examine the Question
   ♦ Recall What You Know
   ♦ Apply What You Know

   ES: C
   G3.4

4. A student conducts an experiment. She puts a different type of soil into each of four identical pots. Each pot has a hole at the bottom. She then puts the same amount of water in each pot. A different amount of water drains out of each pot. What is the best conclusion that can be reached from this experiment?

   A. Different soils have different colors.
   B. Different soils have different textures.
   C. Different soils have different nutrients to support life.
   D. Different soils have different abilities to absorb water.

   ES: C
   G3.5

5. The police arrested a person they suspected of a crime. The police scraped soil from the suspect's shoes for evidence that he was at a certain location. Why did the police decide to scrape his shoes?

   A. Footprints are often left in soft dirt.
   B. Each shoe leaves a unique footprint.
   C. Shoes react differently to different soils.
   D. Soils differ from place to place in color and texture.

   ES: C
   G3.6

6. Jared loves to collect different rocks. One of the rocks in his collection is a dark, heavy piece of granite. It has many large crystals. How did this rock probably form?

   A. from a coral reef
   B   from cooled magma
   C. from the action of glaciers
   D. from pressed layers of sand

   ES: C
   G3.1

7. Which resource is under the wrong heading?

Renewable Resources	Non-Renewable Resources
trees	oil
soil	fresh water

   A. trees
   B. soil
   C. oil
   D. fresh water

   ES: C
   G5.6

LESSON 17

WEATHER

Do you feel like playing outside after school? Is it warm or cold? Is it sunny, raining, or snowing? The weather is important to you since it affects your everyday life. In this lesson, you will learn about weather.

— MAJOR IDEAS —

A. Earth is surrounded by a blanket of air, known as the **atmosphere**. Air takes up space and moves around us as wind. Air pressure can be measured with an instrument called a **barometer**.

B. **Water** exists in the air in different forms. Through the **water cycle**, water passes back and forth between the atmosphere and Earth's surface.

C. **Weather** describes conditions in Earth's atmosphere. It can be measured based on temperature, wind speed and direction, precipitation and barometric pressure. Weather calendars and maps record information about the weather.

D. Different types of clouds usually accompany different kinds of weather. There are four types of clouds you should know: **cirrus**, **cumulus**, **stratus**, and **cumulonimbus**.

E. In the United States, weather patterns generally move from west to east.

THE ATMOSPHERE

When we say it is going to rain, or that it is cold outside, what is the "it" we are talking about? The "it" refers to the **atmosphere** — the blanket of air that surrounds Earth.

You already learned something about the atmosphere in the last lesson. You might recall that the atmosphere is mainly made up of nitrogen and oxygen gas. It is an important renewable resource. The atmosphere is also the source of the weather. In fact, the **weather** really refers to conditions in Earth's atmosphere.

WHAT IS AIR?

The mixed gases in the atmosphere, known as air, are usually invisible to us. However, air is a form of matter. Like every form of matter, air has mass and takes up space. Because we are surrounded by invisible air, we are not always aware of it. But if you put air into a balloon and squeezed it, you would see that the air takes up space inside the balloon. When the air moves around us, we can feel it as **wind**.

AIR PRESSURE

Most of the time, we are not aware of the air, but air is always pushing down on us. The weight of the air causes **air pressure** — the force with which the air presses down. Air pressure is not always the same. It changes based on the amount of moisture in the air, how cold or warm the air is, and other factors. Scientists use an instrument called a **barometer** to measure air pressure. Barometers generally fall into two main groups:

★ Some barometers use mercury and act almost like a mercury thermometer. A glass tube sits in mercury open to the air. Air pressure pushes the mercury up the tube. Air pressure is often measured in **inches of mercury**.

★ A second type of barometer is usually round in shape. It uses a metal spring, like the spring of a spring scale. Air pressure squeezes the spring. A needle points to the amount of air pressure.

A barometer is a very effective tool to let us know what type of weather is ahead. For example, when the barometer is rising, it means that you can expect mild weather — cooler temperatures and clear skies. However, when the barometer has falling air pressure, it indicates that warmer weather, storms or rain is coming.

APPLYING WHAT YOU HAVE LEARNED

◆ What does a barometer measure? _____

◆ How does a barometer help predict the weather? _____

_____

THE WATER CYCLE

Another important factor affecting weather is the amount of water in the air. The **water cycle** is the process by which Earth's water moves into and out of the atmosphere. The water cycle begins when energy from the sun heats the surface of lakes and oceans. This solar energy causes some of this water to **evaporate**. The liquid turns into an invisible gas, **water vapor**, and rises into the atmosphere. In the atmosphere, the water vapor cools and **condenses**, or turns back into tiny droplets of liquid water. These droplets form around particles of dust in the air. They are so light that they still remain in the atmosphere, where they form **clouds**. When the droplets get larger, they fall back to the ground as raindrops, snowflakes, sleet or hail (*precipitation*).

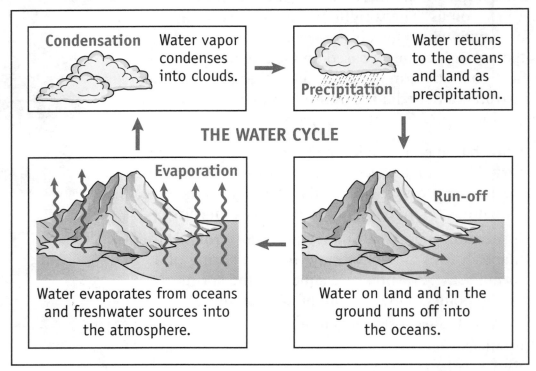

- ★ **Rain** forms when cloud droplets collect and fall through warm air.
- ★ **Snow** forms when water in the air condenses and freezes around ice crystals.
- ★ **Hail** forms when water condenses and freezes around ice crystals in layers, making small balls of ice.

Gravity pulls this water and ice back down to Earth's surface. Some of this water falls on land, where it forms lakes, streams and rivers. Some of the water is absorbed into the ground. It sinks until it hits dense rock and collects as **groundwater**. Much of this water gradually drains back into the ocean. The process then begins all over again as surface water evaporates into the atmosphere. Plants and animals create water vapor as well.

APPLYING WHAT YOU HAVE LEARNED

✦ Describe how the water cycle affects Earth. _____

_____

_____

WEATHER AND CLOUDS

To describe the weather, scientists take a number of key measurements:

★ **Temperature.** Scientists measure how hot or cold the air is.

★ **Precipitation.** Scientists measure the amount of rain, snow, hail, or sleet that falls in inches or centimeters.

★ **Wind.** Scientists measure the speed and direction of the wind.

★ **Barometric Pressure.** Scientists measure air pressure and see if it is rising or falling. This helps them to know what kind of weather is ahead.

CLOUDS

Clouds are formed by ice and water droplets that have condensed in the atmosphere. They can often be used to help predict the weather.

Cirrus Clouds

★ **Cirrus Clouds.** These clouds form out of ice crystals high in the sky. Cirrus clouds tend to look feathery or as a string of clouds — appearing thin and white. They usually indicate good weather.

★ **Cumulus Clouds.** The word *cumulus* comes from the Latin word for a heap or pile. These clouds can form anywhere. They usually appear as puffy white clouds that look like lumpy tops or large cotton balls with flat bottoms. Cumulus clouds also usually indicate good weather.

Cumulus Clouds

★ **Stratus Clouds.** These clouds usually form low in the sky. They often appear as white blankets or layers and can cover the entire sky. Stratus clouds often indicate a gray, dull day with rain. When they appear just above the ground, they cause **fog**.

Stratus Clouds

★ **Cumulonimbus Clouds.** These clouds form when moist, warm air quickly rises. They often appear as white blankets or mushrooms that reach far upwards. *Nimbus* comes from the word for rain. These clouds often bring heavy rain, thunderstorms, and lightning.

Cumulonimbus Clouds

APPLYING WHAT YOU HAVE LEARNED

◆ Based on these cloud formations, what kind of weather would you predict?

A. _____

B. _____

RECORDING WEATHER INFORMATION

Scientists keep track of the weather by recording weather information on a calendar or map. They record the temperature, the direction and speed of the wind, barometric pressure, and precipitation (*rain, snow, and hail*). Scientists use special symbols to record this information.

Cloud Cover	Precipitation	Weather Station Information
○ Clear	•• Heavy rain	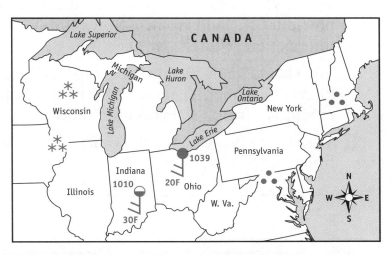
● Overcast	✳✳ Heavy snow	
◐ Partly cloudy	≡ Fog	

WEATHER MAPS

A **map** is a diagram representing a place that shows where something is located. A map's **key** or **legend** explains the symbols used on the map. A **direction indicator** or **compass** shows directions (N, S, E, W). Sometimes a scale is used to tell what distances on the map represent (1 inch = 100 miles).

A **weather map** is a special kind of map showing weather patterns. Special lines may indicate large bodies of cold or warm air. Cloud cover, precipitation, and weather station information may also be found on a weather map. For example, this weather map shows that in Ohio it is 20° Fahrenheit, with 1039 air pressure. It is overcast and the winds are from the south. Can you tell what the weather is in Indiana?

Scientists have found that there are some basic patterns in the weather. That is because the way that the sun heats some areas is usually the same. Areas are also influenced by their nearness to mountains or water. In the United States, weather patterns usually move from west to east. For example, rain in the west moves eastwards. This is often the case because our winds mainly blow from west to east.

APPLYING WHAT YOU HAVE LEARNED

✦ What causes weather patterns in the United States to move from west to east? _____

_____

WHAT YOU SHOULD KNOW

- You should know that Earth is surrounded by a blanket of air, known as the **atmosphere**. Air takes up space and moves as wind. Air pressure can be measured with an instrument called a **barometer**.

- You should know that **water** exists in the air in different forms. In the **water cycle**, water **evaporates** from the sea and land, condenses into clouds in the atmosphere, and falls back to Earth as **precipitation**.

- You should know that **weather** describes conditions in Earth's atmosphere. Weather can be measured based on temperature, wind speed and direction, precipitation and barometric pressure. Weather calendars and maps record information about the weather.

- You should know that different types of clouds usually accompany different kinds of weather. There are four types of clouds you should know: **cirrus**, **cumulus**, **stratus** and **cumulonimbus**.

- You should know that in the United States, weather patterns generally move from west to east.

CHAPTER STUDY CARDS

Cloud Types

Different types of clouds accompany different kinds of weather:

- ★ **Cirrus.** High in the atmosphere; look feathery; indicate fair weather.
- ★ **Cumulus.** Puffy clouds; indicate fair weather.
- ★ **Stratus.** Layers, like a blanket; indicate gray sky and probably rain.
- ★ **Cumulonimbus.** White blanket or mushroom-shaped; bring heavy rain, thunderstorms, and lightning.

Weather

- ★ Indicates the conditions present in the atmosphere.
- ★ Consists of **barometric pressure** (air pressure) temperature, wind speed and direction, and precipitation.

Water Cycle

Water cycle is the process by which water moves into and out of the atmosphere.

- ★ **Evaporation.** The heat of the sun causes water to evaporate and rise in the air.
- ★ **Condensation.** In the atmosphere, water vapor turns back into tiny drops of water.
- ★ **Precipitation.** When droplets get larger, they fall back to Earth as water.

CHECKING YOUR UNDERSTANDING

1. Which part of the water cycle came before the X in the diagram?

ES: D
G4.2

HINT

To answer this question, you need to understand the water cycle: water evaporates into the atmosphere where it forms droplets. These droplets make clouds. Before the water was in the cloud in the atmosphere, it was evaporating from Earth's surface. The best answer is **Choice D**.

Now try answering some additional questions on your own:

2. A student looks outside her window after it has rained and sees puddles of water. Later that same day, the puddles are gone. What process explains why the puddles have disappeared?

A. runoff
B. evaporation
C. precipitation
D. condensation

♦ **Examine the Question**
♦ **Recall What You Know**
♦ **Apply What You Know**

ES: D
G4.3

3. Which of the following is an example of condensation in the water cycle?

A. Streams flow into rivers.
B. Clouds form in the atmosphere.
C. Water drops fall through the air.
D. Puddles disappear on a hot day.

ES: D
G4.3

4. What type of instrument do scientists use to measure air pressure?

A. telescope
B. rain gauge
C. barometer
D. thermometer

ES: D
G4.1

5. In which part of the diagram on the right is water changing from a liquid to a gas?

   A. A
   B. B
   C. C
   D. D

   ES: D
   G4.3

6. Which symbol on a weather map would indicate that the weather is gray and overcast?

   A. ○
   B. ●

   C. ✳
   D. ⦂⦁

   ES: D
   G4.5

7. A student observes several cumulonimbus clouds in the sky. What is the weather most likely to be?

   A. foggy
   B. light rain

   C. a thunderstorm
   D. sunny and warm

   ES: D
   G4.5

8. When scientists observe weather patterns in the United States, in which direction do they find these patterns generally move?

   A. from east to west
   B. from west to east
   C. from north to south
   D. from south to north

   ES: D
   G4.6

9. Which statement about the blanket of air that surrounds us is inaccurate?

   A. Air is the main source of weather on Earth.
   B. When air moves around us, we experience it as wind.
   C. Air is measured by determining its barometric pressure.
   D. The amount of water in the air always remains the same.

   ES: D
   G4.1

10. A barometer is an important instrument for predicting future weather.

    ES: D
    G4.1

    In your **Answer Document**, describe the type of weather that can be expected when the barometer rises and when it falls. (2 points)

11. The water cycle plays an important role in determining the weather.

    In your **Answer Document**, identify two processes in the water cycle. Then explain how each process you selected can have an impact on the weather. (4 points)

    ES: D
    G4.1

CONCEPT MAP OF EARTH AND SPACE SCIENCES

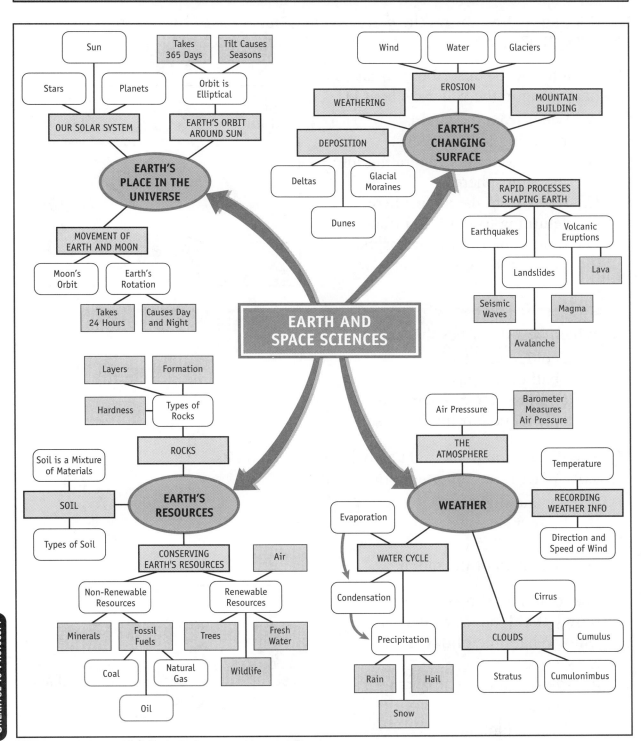

TESTING YOUR UNDERSTANDING

The diagram below shows the water cycle. Four parts of the diagram have been labeled A, B, C, and D.

1. Which best describes what is happening in the water cycle at D?

   A. Water from the land is returning to the oceans.   **ES: D G4.3**

   B. Ocean water is evaporating into the atmosphere.

   C. Water from the atmosphere is returning to Earth's surface.

   D. Water vapor in the atmosphere is condensing into clouds.

WATER CYCLE

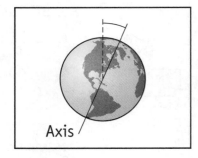

2. Which best describes the sun?

   A. a cold, dead star
   B. a ball of hot gases
   C. a lifeless, rocky environment
   D. a planet collapsing into itself

   ES: A G5.4

3. The illustration to the right shows Earth tilted on its axis. What is an important effect of this tilt?

   A. day and night   **ES: A G5.3**
   B. the four seasons
   C. a year of 365 days
   D. the phases of the moon

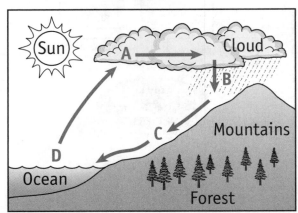

4. The diagram shows a tree on a snowy field on a winter day in Ohio. What time of day is it?

   A. 8:00 A.M.   **ES: A G5.1**
   B. 12:00 noon
   C. 4:00 P.M.
   D. 12:00 midnight

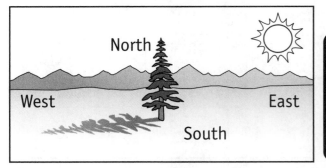

5. Julia is visiting the "Big Island" of Hawaii. She is surprised to see mountains with craters. Inside the craters is hot, molten rock. What are these landforms called?

   A. deltas
   B. dunes

   C. moraines
   D. volcanoes

   ES: B
   G4.8

6. Wave action against solid rock can cause changes in its structure. What is the correct sequence in the erosion of the rock surface shown in the diagrams?

   A. $1 \rightarrow 2 \rightarrow 3 \rightarrow 4$
   B. $3 \rightarrow 4 \rightarrow 2 \rightarrow 1$
   C. $4 \rightarrow 2 \rightarrow 3 \rightarrow 1$
   D. $2 \rightarrow 3 \rightarrow 4 \rightarrow 1$

   ES: B
   G4.8

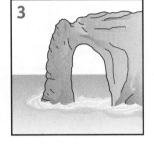

♦ **Examine the Question**
♦ **Recall What You Know**
♦ **Apply What You Know**

7. The surface of Earth can be changed by the action of a moving glacier. Which change would most likely be caused by glacial movement?

   A. creating a delta
   B. causing volcanic eruptions

   C. building up mountains
   D. leaving behind a moraine

   ES: B
   G4.8

8. Students took a field trip to their local courthouse. While there, they noticed small holes in the marble steps of the courthouse. What is the most likely source of this erosion?

   A. winds blowing sand
   B. the rays of the sun

   C. glacial movement
   D. plant roots

   ES: B
   G4.10

9. Which of these is a renewable resource?

   A. gold
   B. wood

   C. coal
   D. petroleum

   ES: C
   G5.6

10. A student mixes together sand, clay, and decayed material from dead plants. What would this mixture be useful for learning about?

   A. rocks
   B. gases

   C. soils
   D. fossils

   ES: C
   G3.4

Use the chart below to answer question 11.

SOIL AND WATER ABSORPTION RATES

Soil Type	Time for 1 Inch of Water to Drip Through
Sand	0.5 hours
Loam	2.0 hours
Silt	2.25 hours
Clay	5.0 hours

11. A scientist tested four types of soils to find which type was best at holding water. The results of the investigation are recorded in the table above. According to the table, which soil type was best at retaining water?

    A. sand
    B. loam
    C. silt
    D. clay

> ♦ **Examine the Question**
> ♦ **Recall What You Know**
> ♦ **Apply What You Know**

ES: C
G3.5

12. Where does petroleum come from?

    A. minerals found in Earth's surface
    B. chemical reactions that took place recently
    C. volcanic ash that mixed with ocean water
    D. remains of organisms living millions of years ago

ES: C
G5.5

ES: C
G3.2

13. Rocks formed in areas once covered by oceans often contain the fossils of animals that lived in the sea. Which of these rocks was once covered by ocean waters?

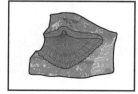  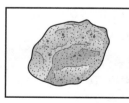 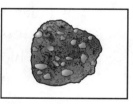

    A.   B.   C.   D.

14. In which of these ways do volcanoes help to build up the Earth's surface?

    A. They add heat to Earth's surface.
    B. They add lava to Earth's surface.
    C. They add gases to the atmosphere.
    D. They add water vapor to the atmosphere.

ES: B
G4.10

15. Earth's water cycle depends on a major source of energy. What supplies the energy that drives the water cycle?

    A. the sun
    B. the moon
    C. the oceans
    D. the eight planets

    ES: D
    G4.3

16. Many different natural processes help to shape Earth's surface.

    In your **Answer Document**, identify two of the processes that help to shape Earth's surface. (2 points)

    ES: B
    G4.10

17. There are many different types of clouds.

    In your **Answer Document**, identify two types of clouds.

    Then, describe the kind of weather each cloud type is likely to accompany. (4 points)

    ES: D
    G4.7

CHECKLIST OF SCIENCE BENCHMARKS

Directions. Now that you have completed this unit, place a check (✔) next to those benchmarks you understand. If you are having trouble recalling information about any of these benchmarks, review the lesson indicated in the brackets.

EARTH SCIENCES

☐ You should be able to explain the characteristics, cycles and patterns involving Earth and its place in the solar system. [**Lesson 14**]

☐ You should be able to summarize the processes that shape Earth's surface and describe evidence of those processes. [**Lesson 15**]

☐ You should be able to describe Earth's resources, including rocks, soil, water, air, animals and plants, and the ways in which they can be conserved. [**Lesson 16**]

☐ You should be able to analyze weather and changes in weather that occur over time. [**Lesson 17**]

A PRACTICE GRADE 5
SCIENCE ACHIEVEMENT TEST

UNIT 6

This unit consists of a complete practice **Grade 5 Science Achievement Test**. Before you begin, let's review a few suggestions to keep in mind for the test:

★ **Answer All Questions.** This practice test consists of 32 multiple-choice questions, four short-answer, and two extended-response questions.

★ **Use the "E-R-A" Approach.** Remember to carefully **examine** the question to understand what it is asking. Next, **recall** what you have learned about that particular topic in science. Finally, **apply** your knowledge to answer the question.

★ **Use the Process of Elimination.** When answering a multiple-choice question, it may be clear to you that certain choices are wrong. After you eliminate incorrect choices, select the **best** response that remains. Never leave a question unanswered, since there is no penalty for guessing. Blank answers are always counted as wrong.

★ **Revisit Difficult Questions.** If you run into a difficult question, do not be discouraged. Put a check (✔) next to it. Answer it as best you can and move on to the next question. At the end of the test, go back and reread any questions you have marked. Sometimes the answer to a difficult question might become clearer to you later.

★ **When You Finish.** When you are finished, check over your work during any time you have left. Do not disturb other students.

Like every question in this book, this practice test identifies the Ohio **standard**, **benchmark**, and **grade level indicator** tested by that question. This will help you to identify any topics you may still need to study further.

Good luck on this practice test!

LESSON 18

A PRACTICE SCIENCE ACHIEVEMENT TEST

Name _____   Date _____

1. Ⓐ Ⓑ Ⓒ Ⓓ
2. Ⓐ Ⓑ Ⓒ Ⓓ
3. Ⓐ Ⓑ Ⓒ Ⓓ
4. Ⓐ Ⓑ Ⓒ Ⓓ
5. Ⓐ Ⓑ Ⓒ Ⓓ
6. Ⓐ Ⓑ Ⓒ Ⓓ
7. Write your response to question 7 in your answer document.
8. Ⓐ Ⓑ Ⓒ Ⓓ
9. Ⓐ Ⓑ Ⓒ Ⓓ
10. Ⓐ Ⓑ Ⓒ Ⓓ
11. Ⓐ Ⓑ Ⓒ Ⓓ
12. Write your response to question 12 in your answer document.
13. Ⓐ Ⓑ Ⓒ Ⓓ
14. Ⓐ Ⓑ Ⓒ Ⓓ
15. Ⓐ Ⓑ Ⓒ Ⓓ
16. Ⓐ Ⓑ Ⓒ Ⓓ
17. Ⓐ Ⓑ Ⓒ Ⓓ
18. Ⓐ Ⓑ Ⓒ Ⓓ
19. Ⓐ Ⓑ Ⓒ Ⓓ
20. Ⓐ Ⓑ Ⓒ Ⓓ

21. Write your response to question 21 in your answer document.
22. Ⓐ Ⓑ Ⓒ Ⓓ
23. Ⓐ Ⓑ Ⓒ Ⓓ
24. Ⓐ Ⓑ Ⓒ Ⓓ
25. Ⓐ Ⓑ Ⓒ Ⓓ
26. Ⓐ Ⓑ Ⓒ Ⓓ
27. Write your response to question 27 in your answer document.
28. Ⓐ Ⓑ Ⓒ Ⓓ
29. Ⓐ Ⓑ Ⓒ Ⓓ
30. Ⓐ Ⓑ Ⓒ Ⓓ
31. Ⓐ Ⓑ Ⓒ Ⓓ
32. Ⓐ Ⓑ Ⓒ Ⓓ
33. Ⓐ Ⓑ Ⓒ Ⓓ
34. Write your response to question 34 in your answer document.
35. Ⓐ Ⓑ Ⓒ Ⓓ
36. Ⓐ Ⓑ Ⓒ Ⓓ
37. Write your response to question 37 in your answer document.
38. Ⓐ Ⓑ Ⓒ Ⓓ

Use the following diagram to answer questions 1–3.

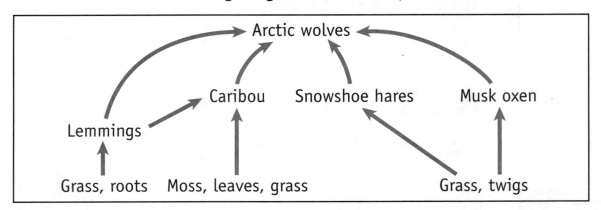

1. What do the arrows in this diagram represent?

   B

   A. chemical changes

   B. the transfer of energy

   C. the life cycles of animals

   D. the extinction of some plants and animals

   LS: B
   G5.1

2. Which animal in this food web is an omnivore?

   A

   A. caribou

   B. lemming

   C. arctic wolf

   D. snowshoe hare

   LS: B
   G5.3

3. Based on the food web above, what would most likely happen if the number of arctic wolves increased?

   C

   A. The amount of grass would decrease.

   B. The number of caribou would increase.

   C. The number of musk oxen would decrease.

   D   The number of snowshoe hares would increase.

   LS: C
   G5.5

4. Which is an example of a physical change?

   B

   A. A piece of paper burns.

   B. A steel rod is heated until it melts.

   C. Hydrogen and oxygen gases combine to make water.

   D. A plant turns water and carbon dioxide into sugar and oxygen.

   PS: A
   G4.1

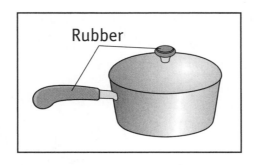

Rubber

5. Why do manufacturers often put rubber handles on metal pots?

   A. Both rubber and metal conduct heat well.

   B. Neither metal nor rubber conduct heat well.

   C. Rubber does not conduct heat as well as metal does.

   D. Metal does not conduct heat as well as rubber does.

PS: B
G4.3

6. The illustration to the right shows four different parts of a plant. Which part of the plant transports water from the roots to the leaves?

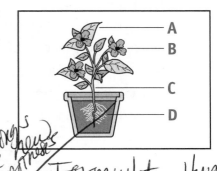

   A. (A)

   B. (B)

   C. (C)

   D. (D)

LS: B
G4.2

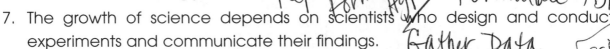

7. The growth of science depends on scientists who design and conduct experiments and communicate their findings.

In your **Answer Document**, explain two steps scientists take in designing an experiment, and two steps scientists take in conducting an experiment. (4 points)

SI: C
G4.3

8. Two students wanted to find out which of their toy trucks would go farthest. They decided to let each truck roll down a ramp and then measure how far it rolled on the ground. What condition must be held constant if they want a fair test?

   A. the time of day

   B. the incline of the ramp

   C. the weight of the ramp

   D. the temperature of the room

SI: C
G4.4

9. Based on the weather map to the right, what will be tomorrow's weather in Columbus?

A. cold and dry
B. cool and rainy
C. cold and snowy
D. warm and sunny

ES: D
G4.5

Pg. 195

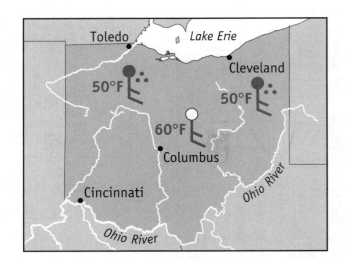

10. A student conducted an experiment to find out how temperature affects the volume of air in a balloon. He drew a line around the center of a blown-up balloon. Next, he put the balloon in the freezer. Later he took it out and measured the length of the line. The results are recorded in the chart. What conclusion can be drawn from these results?

HOW TEMPERATURE AFFECTS AIR IN A BALLOON

Conditions of Balloon	Length of Line Around Balloon (in centimeters)
Balloon after coming out of the freezer	14 cm
Balloon at room temperature	21 cm
Balloon after being warmed for 2 min	33 cm
Balloon after being warmed for 4 min	54 cm

A. The balloon is larger in the freezer than outdoors.
B. The balloon is unaffected by changes in temperature.
C. The warmer the balloon gets, the larger it becomes.
D. The warmer the balloon gets, the smaller it becomes.

SI: B
G3.3

11. What characteristic of Earth is shared by the other planets of the solar system?

A. They have gravity.
B. They have a layer of soil.
C. Much of their surface is covered by water.
D. They are surrounded by a thin blanket of air.

ES: A
G5.3

12. Scientists often sort objects based on their physical properties.

    In your **Answer Document**, identify two physical properties of a piece of iron. (2 points) *Magnetic – strong – sturdy*

    PS: B G4.3

    Use the following diagram to answer questions 13–15.

13. The illustration above shows a tropical rainforest. What gas in the atmosphere would increase if people cut down the trees in this rainforest?

    A. oxygen   C. water vapor

    B. nitrogen   D. carbon dioxide   *D*

    LS: C G5.6

14. What happens to the dead leaves and animals that fall to the forest floor?

    A. They disintegrate into the atmosphere.

    B. They remain preserved in the rainforest.   *C*

    C. They decay and become part of the soil.

    D. They are washed into the oceans by heavy rains.

    ES: C G3.4

15. How do the trees in this rainforest obtain their energy?

    A. by eating other plant life

    B. by using energy from the sun

    C. by drinking water through their roots   *B*

    D. by absorbing heat from the forest floor

    LS: B G5.1

16. What step can farmers take to prevent soil erosion from heavy rains?

   A. planting grass
   B. cutting trees

   C. spraying with water
   D. raising grazing animals

   ES: C
   G5.6

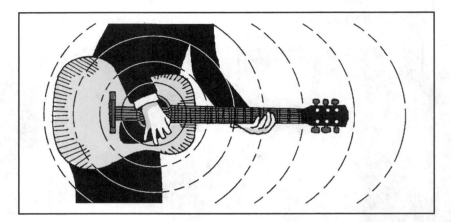

17. A student plucks at the strings of his guitar. Why do some of the strings make a higher-pitched sound than other strings?

   A. They vibrate faster than others.
   B. They produce fewer sound waves.
   C. Their vibrations pass through less air.
   D. They are closer to the neck of the guitar.

PS: F
G5.7

18. Which instrument can be used to measure air pressure?

ES: D
G4.1

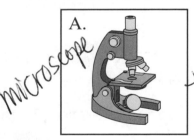   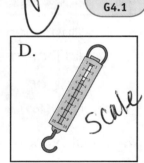

microscope   thermometer   barometer   scale

19. A scientist wants to determine what food a specific type of butterfly usually eats. What would be the best way to find an answer to this question?

   A. Observe the butterfly in its natural surroundings.

   SW: B
   G5.4

   B. Conduct an experiment by giving the butterfly different foods.
   C. Ask friends what they think the butterfly eats.
   D. Create a model of this type of butterfly.

20. Which diagram shows a process that builds up Earth's landforms?

ES: B
G4.10

A.   B.   C.   D.

21. Students are conducting an experiment to see how thermal energy can be changed or transferred. *Heat from a stove or candle*

    In your **Answer Document**, identify one way in which the thermal energy of an object might be changed. *Melting chocolate*

    PS: D
    G5.1

    Then identify one instrument that is used by scientists to measure thermal energy. (2 points) *A thermometer*

Chocolate Chips

Use the following illustration to answer question 22.

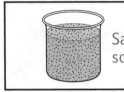

Sandy soil	Soil with high clay content	Soil with high content of decayed plant material
1	2	3

22. Students collected the three samples of different types of soils shown above. They inspected the three samples with a hand lens and rubbed each sample with their hands. What question was this investigation designed to answer?

    A. What is the texture of each type of soil?
    B. What microscopic bacteria live in the soil?
    C. How well does each type of soil hold water?
    D. How well does each type of soil support life?

    SW: B
    G5.4

23. What causes Earth's changing seasons?

    A. the moon's orbit around Earth
    B. Earth's spinning around its axis
    C. the effect of nearby planets on Earth
    D. Earth's tilt as it revolves around the sun

    ES: A
    G5.3

24. Tiny drops of water are light enough to float in the air as clouds. As these drops of water bump into each other, they combine into larger drops of water. What happens when these drops become too large to float in the air?

    A. They fall as rain.
    B. They create fog.
    C. They evaporate.
    D. They collect underground.

    ES: D
    G4.2

25. A group of students observe cars stopping on a rough gravel road. Then they observe other cars stopping on a smooth, paved roadway.

    What conclusion are they most likely to draw from their observations?

    A. Both roads have the same friction.
    B. The paved road has greater friction.
    C. The gravel road has greater friction.
    D. At different speeds, each road has greater friction.

    PS: C
    G3.4

Use the following photographs to answer question 26.

A river cuts through a canyon

A landslide

26. How are the erosion of rock by running water and a landslide different?

    A. Water erosion is a slower process.
    B. Landslides can affect human activity.
    C. Water erosion changes Earth's surface.
    D. Landslides move materials to a new place.

    ES: B
    G4.10

27. Modern technology has had both positive and negative effects on human life.

    In your **Answer Document**, identify one beneficial effect and one harmful effect of modern technology. (2 points)

    ST: A
    G3.2

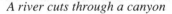

More jobs | helps economy | More pollution / CO2

Use the following illustration to answer question 28.

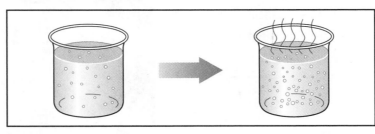

28. The diagram above shows water in a container. What process is taking place?

   A. boiling
   B. melting

   C. freezing
   D. condensation

Renewable Resources	Non-Renewable Resources
Fresh water	Copper
Trees	? *Coal*

29. Which of these best completes the above chart?

   A. coal
   B. salmon

   C. soil
   D. wind energy

1	2	3 
4 ✓ 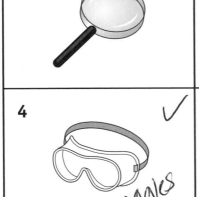 *goggles*	5 	6 ↓  *gloves*

30. Which pieces of laboratory equipment shown above are used for safety?

   A. 1 and 2
   B. 3 and 5

   C. 3 and 4
   D. 4 and 6

31. The Australian echidna, or spiny anteater, lives in the desert and eats ants. Which structure most helps the echidna obtain its food?

A. short legs

B. sharp spines

C. poor eyesight

D. long, sticky tongue

LS: B
G3.2

32. While playing basketball, a student throws a ball into the air. What force brings the basketball back to the ground?

A. gravity

B. friction

C. magnetism

D. conductivity

PS: C
G3.3

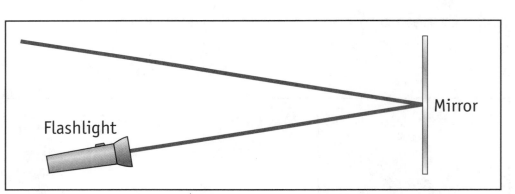

Flashlight

Mirror

33. What happens when light strikes a smooth, shiny surface like a mirror?

A. The light is refracted.

B. The light is absorbed.

C. The light is reflected.

D. The light is transmitted.

PS: F
G5.5

34. When wind, water or ice erodes rock or soil in one area, it deposits these materials in a different area.

In your **Answer Document**, identify two types of landforms often produced by this process of deposition. (2 points)

ES: B
G4.8

35. A graduated cylinder contains 25 mL of water. A pebble is then placed in the cylinder. The diagram shows the volume of water after the pebble is placed in the cylinder. What is the volume of the pebble?

   A. 5 cm³
   B. 15 cm³
   C. 20 cm³
   D. 55 cm³

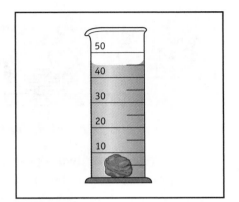

SI: A G4.1

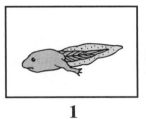 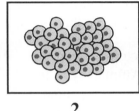  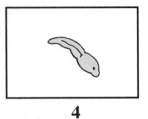

1   **2**   **3**   **4**

36. Which is the correct order of development for this frog?

   A. 4, 2, 3, 1   C. 2, 1, 4, 3
   B. 3, 1, 4, 2   D. 2, 4, 1, 3

LS: A G3.1

37. Plants and animals have different structures to help them survive and reproduce.   *nectar to attract bees — spreads pollen*
   In your **Answer Document**, identify one plant structure and describe its function.   *Pg. 86*
   Then identify one animal structure and describe its function. (4 points)   *pg. 97*

LS: B G4.2

38. A wire is wrapped around a piece of iron. When both ends of the wire are connected to a battery, the paper clips are attracted to the coiled metal. Why are the paper clips attracted?

   A. The paper clips become electric.
   B. Gravity pulls the objects together.
   C. Thermal energy heats up the metal.
   D. Electricity makes the rod magnetic.

PS: E G5.3

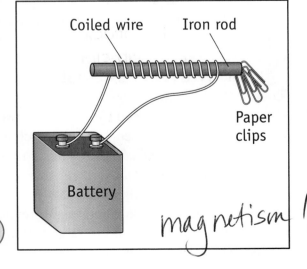

Coiled wire   Iron rod

Paper clips

Battery

magnetism!

The number in brackets indicates the page number where the term is first discussed.

Air Pressure. Air is always pushing down on us. The weight of the air causes the force with which the air pushes down, known as air pressure. [191]

Animal. A living thing that can move freely but cannot produce its own food. An animal eats plants or other animals to survive. [95]

Atmosphere. The blanket of air surrounding Earth. [190]

Avalanche. When snow, ice, and rock fall quickly down the side of a mountain. [177]

Backbone. Characteristic used by scientists to classify animals. Animals without a backbone include jellyfish, ants, snails, and lobsters. Animals with backbones can be divided into five main groups: fish, amphibians, birds, reptiles, and mammals [97]

Balance. A dual-pan or triple-beam balance is used to measure mass. [48]

Barometer. An instrument used by scientists to measure air pressure. [191]

Carnivore. An animal, such as a lion, that only eats other animals. [109]

Chemical Change. When a substance changes its structure. For example, when a log burns in a fireplace it changes from wood into a heap of ashes. [132]

Chemical Property. The ability of one material to combine with others in a chemical reaction. Hydrogen, for example, can burn. Water cannot. [133]

Cirrus Clouds. Clouds formed out of ice crystals high in the sky. They tend to look feathery, appearing thin and white. They usually indicate good weather. [193]

Collision. This occurs when one object crashes into or collides with another object. [140]

Competition. When living things in an ecosystem struggle with others for the same resources. They compete for food, water, and space. [107]

Conductivity. The ability of matter to carry heat, sound, or electricity. [126]

Consumers. Living things that do not make their own food; animals. [109]

Contact Forces. Collisions and friction are both contact forces. [139]

Controlled Experiment. When a scientist designs special tests that are conducted in a closed laboratory. Usually the scientist changes just one variable and measures the effect. [28]

Cooperation. Living things in the same environment help each other to survive. [107]

Clouds. These are formed by ice or tiny water droplets that have condensed in the atmosphere. They can often be used to help predict the weather. [193]

Cumulonimbus Clouds. These clouds form when moist, warm air quickly rises. They often appear as white blankets or mushrooms that reach far upwards. These clouds often bring heavy rain, thunderstorms, and lightning. [194]

Cumulus Clouds. These clouds can form anywhere. They usually appear as puffy white clouds that look like lumpy tops or large cotton balls with flat bottoms. Cumulus clouds also usually indicate good weather. [193]

Day / Night. Cycle of light and darkness occurring on Earth every 24 hours: caused by Earth's rotation on its axis. [166]

Decomposers. Living things in an ecosystem, like ants, worms, and bacteria, which live by breaking down waste products and dead things. [109]

Delta. An area that forms where a river flows into an ocean, sea, or lake. The river carries soil and other sediment. The sediment is deposited at its mouth. It gradually builds up, extending the land in a typical triangle shape. [174]

Deposition. The process by which the rocks, soil, and other sediment moved by erosion are left or deposited in a new place. This creates typical landforms, such as dunes, deltas, and glacial moraines. [173]

Deserts. Regions, such as the Sahara Desert, that receive less than 10 inches of rainfall each year. [106]

Direction. The path or route that an object takes when it moves. [138]

Dunes. Hills of sand caused by deposition. [174]

Earthquake. Vibrations of Earth's crust that send out seismic waves. [176]

Echo. Reflected sound. Curtains, carpets, and thick fabrics are able to absorb sound. Engineers use these sound characteristics when designing buildings. [153]

Ecosystem. All the living and non-living things in an area; they affect and depend on each other. Energy in an ecosystem flows from producers to consumers. [103]

Electrical Circuit. A closed path that electricity follows. Any break in the path will stop the flow of electricity throughout the entire circuit. [149]

Electricity. A form of energy. It is made by fast-moving particles. Like heat and sound, it can pass through some materials. [127, 149]

Energy. Something with the power to do work. Energy takes different forms. Electricity, light, and thermal energy are all different forms of energy. [147]

Erosion. The process by which rock and soil are worn down and carried away by wind, running water, ocean waves or glaciers. [172]

Evaporation. When a liquid disperses into the air as a gas: for example, water evaporates. [130]

Extinct. A type of plant or animal that no longer exists. [113]

Fact. A statement that can be checked to see if it is correct. [26]

Field Investigation. When a scientist goes out into the natural world to collect data. [28]

Food Chain. A chain showing the flow of energy between different plants and animals in an ecosystem; shows what eats what. [110]

Food Web. A diagram showing the interaction of several food chains. [110]

Fossil. Impression left in sedimentary rock by the remains of a dead plant or animal. [113]

Fossil Fuels. Non-renewable resource made from the remains of living organisms millions of years ago that can now be burned for energy. Oil, coal, and natural gas are fossil fuels. [186]

Force. A push or pull acting on an object that will cause it to move. [139]

Friction. The contact force caused by the rubbing together of two or more surfaces. [141]

Gas. A state of matter without any fixed shape or volume, such as oxygen gas. [130]

Glacier. A giant sheet of slow-moving ice. As glaciers move, they scrape Earth's surface. [174]

Glacial Moraine. Material left behind when a glacier retreats. [173]

Global Warming. Increased carbon dioxide in the atmosphere acts as a blanket, trapping in heat. This has led to higher temperatures across the globe. [185]

Graduated Cylinder. A container used to measure the volume of liquids. [48]

Gravity. A non-contact force of attraction between two objects. Gravity pulls objects to Earth. The larger a planet is, the stronger its force of gravity. [139]

Herbivore. An animal, such as a cow, that eats only plants. [109]

Hypothesis. An educated guess that attempts to answer a question. [44]

Kilometer. A distance of 1,000 meters. [57]

Kilogram. A mass of 1,000 grams. [57]

Landslide. When rocks, soil, and other materials suddenly drop down a steep slope or mountain. [176]

Leaf. Part of a plant that makes food from the sun's energy. [87]

Light. A form of energy that can travel through some materials or through empty space. [151]

Liquid. A state of matter where particles move more quickly than in a solid; a liquid has no fixed shape but still has a fixed volume. [130]

Liter. A measure of the volume of a liquid. [57]

Life Cycle. Changes that a plant or animal undergoes from its beginning to its death. [89]

Magnet. A special piece of iron or other material that can attract other metals such as iron, nickel, and steel. [126]

Magnetism. A non-contact force of attraction between a magnet and many metals. [140]

Mass. The amount of matter something has, measured in grams or kilograms. [125]

Matter. Anything that has mass and takes up space. [124]

Metamorphosis. A process in which an animal completely changes its form during its life cycle. Insects and amphibians often go through a metamorphosis. For example, a frog lays an egg which becomes a tadpole and then a frog. A butterfly lays an egg, and a caterpillar comes out. This caterpillar later wraps itself up in a chrysalis. After a period of time, an adult butterfly emerges from the chrysalis. [98]

Microscope. A laboratory instrument used to magnify small objects. [48]

Mixture. When two substances, like sand and salt, are mixed together. [128]

Model. Something made to represent something else and show how it works. [31]

Moon. Earth's natural satellite. [167]

Mountain-building. Sometimes sections of Earth's crust slowly squeeze together over millions of years. Earth's crust starts to fold, pushing some of Earth's crust upwards and creating mountains. [176]

Motion. When an object changes its position over time. Motion consists of speed and direction. [137]

Non-contact Forces. Gravity and magnetism are both non-contact forces. [139]

Non-renewable Resource. A resource formed over a very long period of time that cannot be re-grown or replaced, such as a fossil fuel (oil or coal). [185]

Omnivore. An animal that eats both plants and animals. [109]

Opinion. A statement of personal feeling or belief. [26]

Orbit. The revolving of one body around another. Earth goes around the sun in an elliptical (oval) orbit in just over 365 days (one year) for Earth to complete one orbit around the sun. [166]

Photosynthesis. The process by which plants capture energy from sunlight and convert this energy into food. [87]

Physical Change. When an object changes one or more of its physical properties. For example, a physical change occurs when a pot of water boils. [131]

Pitch. How high or low a sound appears to us. More vibrations per second create a higher pitch. Fewer vibrations make the pitch lower. [153]

Planets. Large bodies of rock, often with ice or gas, that orbit the sun and dominate their orbits. There are eight planets: Mercury, Venus, Earth, Mars, Jupiter, Saturn, Uranus, and Neptune. [165]

Plant. A living thing that cannot move from place to place, but that produces its own food through photosynthesis. [86]

Predator. An animal that hunts and eats other animals. [107]

Prediction. A statement that tells what will probably happen in the future. [67]

Prey. An animal hunted by a predator. [108]

Producer. A living thing that can make its own food; a plant. [86, 108]

Reflection. The bouncing of light off a shiny surface, such as a mirror. [151]

Refraction. A bending of light when it enters a new material, such as a glass lens. [151]

Renewable Resource. A resource, like a tree, that can be replaced or re-grown. [184]

Rock. A solid made of minerals found on Earth's surface or below it. Rocks can be classified by their texture, hardness, the minerals they contain, and how they were formed. [180]

Roots. The part of a plant below the ground that anchors the plant and absorbs water and nutrients. [87]

Rotation. Earth spins around its axis, an imaginary line running through the center of Earth. This rotation takes 24 hours, causing day and night on Earth. [166]

Safety Goggles. Special plastic glasses that cover the eyes, used to protect them during scientific experiments. [48]

Science. A special way of investigating and explaining what happens in nature. [25]

Scientific Inquiry. Usually begins with observation of the natural world. This is followed by questions about what is observed. Then scientists try to find answers to their questions by designing and conducting experiments for further observations. [42]

Sedimentary Rock. Rock made by layers of sand, mud or other materials that are deposited and pressed together. These rocks usually show their layers. Fossils are often found in sedimentary rock. [181]

Soil. A mixture of sand, clay, and decayed plants and animals. Soil can hold water and is found on top of Earth's land surface. Soil types differ by texture and their ability to support life. [182]

Soil Conservation. To conserve the soil against erosion, farmers plant crops to hold the soil together against the wind or heavy rains. Farmers also plow across hills and slopes instead of up and down to reduce soil erosion from running water. [183]

Solid. A state of matter in which volume and shape are fixed, such as ice. [130]

Sound. This is caused by vibrating objects. The vibrations carry energy that our ears can hear. Sound always travels in waves through the air or some other material. [150, 153]

Spring Scale. A scale that measures the weight of an object by seeing how much it pulls on a steel spring attached to a dial. [48]

Stars. Enormous balls of super-heated gases in space, which appear as points of light from Earth. [164]

Stem. The main body of a plant; the stem brings water and minerals to the leaves and carries food from the leaves to the roots. [87]

Stratus Clouds. Clouds that usually form low in the sky, appear as white blankets or layers, and that often indicate a gray, dull day with rain. [194]

Sun. The star at the center of the solar system. [165]

Technology. The use of tools and techniques for making and doing things. [73]

Technological Design. The first step is to identify a need; then identify possible solutions; next, design a solution; and finally, evaluate and test the solution. [78]

Temperate Forests. These develop in regions with 30 to 60 inches of rain each year. The four seasons are marked by moderate temperatures and cool winters. [105]

Temperature. Measures the thermal energy of an object — how fast its particles are moving. [148]

Theory. A "big idea" in science that tries to explain why things happen. [34]

Thermal Energy. Heat caused by the movement of particles that make up matter. [148]

Thermometer. An instrument used to measure temperature in degrees. Scientists measure temperature in degrees Celsius. [48, 62]

Variable. Anything that can be changed or that might change in an experiment. [45]

Volcano. An opening in Earth's surface that lets out molten rock and gases. [176]

Volume. How much space something takes up. [59]

Water Cycle. A cycle in which water from lakes and oceans is heated by energy from the sun, evaporates into the atmosphere, condenses into drops of water that float as clouds, and then falls back to Earth as precipitation (rain, snow, sleet, or hail). [192]

Weather. Conditions in the atmosphere that change daily. [190]

Weathering. The gradual wearing down of rocks on Earth's surface by the action of the wind, water, ice and living things. [171]

Weather Map. A special kind of map showing weather conditions. [195]

Weight. How heavy something is. Weight is created by the force of gravity, and depends on how strongly gravity pulls an object towards Earth. [60, 125]